Applied Aspects of Modern Metrology

Edited by Oleh Velychko

Published in London, United Kingdom

IntechOpen

Supporting open minds since 2005

Applied Aspects of Modern Metrology
http://dx.doi.org/10.5772/intechopen.95661
Edited by Oleh Velychko

Contributors
Oleh Velychko, Tetyana Gordiyenko, Romain Coulon, Oleksandr Samoilenko, Volodymyr Zaets, Valentyn Isaiev, Iurii Anokhin, Jaromír Volf, Viktor Novák, Vladimír Ryzenko, Stanislava Papežová, Oleg Angelsky, Peter Maksymyak, Claudia Zenkova, Olexander Ushenko, Jun Zheng

First published in London, United Kingdom, 2022 by IntechOpen
IntechOpen is the global imprint of INTECHOPEN LIMITED, registered in England and Wales, registration number: 11086078, 5 Princes Gate Court, London, SW7 2QJ, United Kingdom
Printed in Croatia

British Library Cataloguing-in-Publication Data
A catalogue record for this book is available from the British Library

Additional hard and PDF copies can be obtained from orders@intechopen.com

Applied Aspects of Modern Metrology
Edited by Oleh Velychko
p. cm.
Print ISBN 978-1-80355-048-0
Online ISBN 978-1-80355-049-7
eBook (PDF) ISBN 978-1-80355-050-3

For EU product safety concerns:
IN TECH d.o.o., Prolaz Marije Krucifikse Kozulić 3, 51000 Rijeka, Croatia, info@intechopen.com or visit our website at intechopen.com.

Meet the editor

Prof. Oleh Velychko received a master's degree in Radio Engineering and a Ph.D. in Metrology. Since 2005, he has been Director of the Institute of Electromagnetic Measurements of the State Enterprise "Ukrmetrteststandard," Kyiv, Ukraine. He has experience in creating and comparing national standards of electrical quantities. He is also a committee member for the International Measurement Confederation (IMEKO), the International Electrotechnical Commission (IEC), and the Euro-Asian Cooperation of National Metrological Institutions (COOMET). Prof. Velychko has published several book chapters.

Contents

Preface

In the modern era of scientific and technological development, the role of measurements and metrology in scientific research is becoming more and more important due to the increase in the testing of various products. Moreover, requirements for the accuracy and reliability of measurement results are increasing significantly and their ranges are expanding. Improving measurement accuracy allows us to identify the shortcomings of certain technological processes and either eliminate them or reduce their influence. This leads to better-quality products and contributes to saving energy and other resources, as well as raw materials and materials.

This book discusses relevant aspects of practical metrological activity to establish traceability of measurements while increasing their accuracy and reliability. It also presents procedures for the calibration and testing of measuring instruments.

Chapter 1, “Evaluation of the Long-Term Stability of Metrology Instruments”, emphasizes the issue of the long-term stability of instruments used in metrology. It introduces control charts and tests them using simulated and real data sets from metrology instruments. The chapter also discusses the interest and the limits of such statistical analysis.

Chapter 2, “The Data Evaluation of Interlaboratory Comparisons for Calibration Laboratories”, presents features of the data evaluation of interlaboratory comparisons for calibration laboratories. It also presents the results of this evaluation, which show that for such interlaboratory comparisons it is necessary to evaluate the data using the z index.

Chapter 3, “New Trends of Optical Measurements”, presents some of the achievements in modern optical measurements from the nano to the femto or pico units of the measured quantity. Using polarization-holographic measurements in biomedical applications makes it possible to obtain a reliable diagnosis of pathological states of biological objects.

Chapter 4, “Calibration of Tanks and Ships’ Tanks for Storage and Transportation of Liquids by Laser Scanning”, presents the calibration methodology of tanks and ship tanks for storage and transportation of liquids by laser scanning. The developed methods and algorithms will allow, after several times, to increase the accuracy of determining the interval capacities of the tanks with a significant reduction in time for measurements and processing of results.

Chapter 5, “Properties of Tactile Sensors Based on Resistive Ink and the Dimension of Electrodes”, presents the results of research on selecting suitable electrodes for use in the pressure distribution measurement system Plantograf. The aim of the measurement is to determine the usable working range of pressures and the corresponding sensitivity for certain combinations of electrodes and inks.

Chapter 6, “Methodological Aspects of Using Comparators for Metrological Traceability of Instrument Transformers”, presents methodological aspects of using comparators for metrological traceability of voltage and current instrument transformers. The accuracy of both transformers is a critical parameter in terms of ensuring the reliability of functioning of both high- and low-voltage networks.

My appreciation goes to all authors for their valuable contributions.

Oleh Velychko
State Enterprise “Ukrmetrteststandard”,
Kyiv, Ukraine

Chapter 1

Evaluation of the Long-Term Stability of Metrology Instruments

Romain Coulon

Abstract

This chapter aims to emphasize the issue of the long-term stability of instruments used in metrology. This issue is a concern mentioned in the IEC/ISO17025:2017 standard and the JCGM100:2008 guide. Control charts are mentioned in these key documents as tools to assess whether a measurement process is under statistical control or not. Control charts (Shewhart charts, CUSUM chart, EWMA chart) are introduced and tested with simulated and real datasets from metrology instruments that operate at the ionizing department of the BIPM. The interest and the limits of such statistical analysis are discussed. They take their basis in a measurement model composed of Gaussian white noise. Although a measurement monitored over a relatively short period may be consistent with this model, it has been observed that the autocorrelation of the measurement data acquired over a long period limits the relevance of control charts. In this case, time series analysis seems more appropriate than conventional control charts. As an illustration, an optimal Bayesian smoother is introduced to demonstrate how to deconvolve the low-frequency random noise and refine the evaluation of uncertainty according to the measurement model for long-term measurement.

Keywords: metrology, control charts, time series, statistical control, long-term stability

1. Introduction

Long-term reproducibility in experimental sciences is a critical metrological topic recently discussed in [1]. In laboratories, the quality system is the frame in which such an issue is controlled. It is notably required in the ISO/IEC 17025 [2] that laboratories must ensure the validity of their results. For this, monitoring of the state of operating of a measurement system must be set up by periodic checks and statistical analysis of the obtained process control value.

Techniques to assess the reproducibility of a process is a matter of concern since the industrial era. An important step has been taken in 1924 with the invention of control charts by Walter Andrew Shewhart (Bell Telephone). Control charts are efficient tools to control if a process is under statistical control and is now wildly used in many domains and recognized by the Project Management Institute as one of the seven basic tools of quality [3]. Moreover, control charts are explicitly proposed in ISO/IEC 17025 [2] and the JCGM100:2008 Guide [4] as tools to implement to analyze periodic checks using control standards.

 IntechOpen

The main advantages of control charts are their simplicity of implementation and their generic use, independent of the domains. A control chart displays a set of points and control limits. This representation makes it easy to check the number of points exceeding the control limits. Therefore, control charts can be analyzed by anybody without any specific skills in statistics.

Four typical control charts are introduced in this document: the original Shewhart charts (x-chart and R-chart), the CUSUM chart, and the EWMA chart. The manners that they are introduced below should be considered as examples that can slightly vary from one reference to another [5, 6]. These techniques are applied below to simulated data and real data from metrology instruments used in ionizing radiation metrology. Then, the interest of time series analysis is highlighted for the monitoring of measures over a long period using the same datasets.

2. Introduction to control charts

2.1 The x-chart

Periodical control using a check standard consists of periodically reproduce the measurement x_i at the time t_i and forms the measurement set $(x_1, x_2, \ldots, x_i, \ldots, x_N)$ where x_N is the last recorded point. In the x-chart, a central line is drawn by the moving empirical average $\overline{x}_N$ over a period of 31 samples such as

$$\begin{cases} \overline{x}_N = \dfrac{\sum_{i=1}^{N} x_i}{N}, \text{ for } N \leq 31 \\ \overline{x}_N = \dfrac{\sum_{i=N-31}^{N} x_i}{31}, \text{ for } N > 31. \end{cases} \tag{1}$$

The control limits (LCL and UCL) are given by the multiple of the standard deviations:

$$\begin{cases} s(x_N) = \sqrt{\dfrac{\sum_{i=1}^{N} (x_i - \overline{x}_N)^2}{N-1}}, \text{ for } N \leq 31, \\ s(x_N) = \sqrt{\dfrac{\sum_{i=N-31}^{N} (x_i - \overline{x}_N)^2}{30}}, \text{ for } N > 31, \end{cases} \tag{2}$$

such as

$$\begin{cases} \mathrm{LCL} = \overline{\mathrm{x}}_\mathrm{N} - \mathrm{ks}(\mathrm{x}_\mathrm{N}) \\ \mathrm{UCL} = \overline{\mathrm{x}}_\mathrm{N} + \mathrm{ks}(\mathrm{x}_\mathrm{N}), \end{cases} \tag{3}$$

where the multiplication factor k is set equal to 2 for the first control level and 3 for the second control level.

Failure criteria are defined according to a white Gaussian model that will be introduced later. It is important to note that the relevance of the analysis is low when the number of measures $N < 10$, relevant for $10 \leq N \leq 31$, and good for $N \geq 31$. Some criteria can be founded in the ISO 7870 standard [5]. As an example, the following rules are proposed where a measurement process is not considered under statistical control when:

- more than 0.35% of the points are above or below the control limits with the multiplication factor $k = 3$,
- more than 20% of the points are above or below the control limits with the multiplication factor $k = 2$,
- more than 9 points are in a row one side of $\overline{x}_N$,
- more than 6 points are in a row increasing or decreasing,
- more than 14 points are in a row alternating between increase and decrease.

2.2 The R-chart

In the R-chart, the plotted points R_i are the derivatives of the measurements:

$$R_i = x_i - x_{i-1} \text{ for all } i \in [\![2, N]\!]. \tag{4}$$

The central line and two levels of control limits are obtained in the same manner as for the x-chart by calculating the moving empirical average and standards deviations of the values R_i. Also, the same failure criteria can be applied.

2.3 The CUSUM chart

The cumulative of deviations C_i is obtained by the following recursive formula:

$$C_i = C_{i-1} + x_i - \overline{x}_N \tag{5}$$

In this chart, no central line is displayed but two control limits V_i are defined by the so-called V-mask:

$$V_i = C_N \pm ks(x_i)(d + N - i) \tag{6}$$

where $k = 0.1$ and $d = 147.22$.

Any points out of the V-mask mean that a significant drift of the process control value is detected.

2.4 The EWMA chart

The EWMA chart is obtained by plotting the exponentially smoothed data points z_i by using the following recursive formula:

$$z_i = \lambda x_i + (1 - \lambda) z_{i-1} \tag{7}$$

And where λ is a smoothing parameter.

The control limits CL are defined by

$$CL_i = \overline{x}_i \pm 3\, s(x_i) \sqrt{\frac{\lambda}{2 - \lambda}\left(1 - (1 - \lambda)^{2i}\right)} \tag{8}$$

With the smoothing parameter $\lambda = 2/N$.

Any point out of these control limits means that a significant drift of the process control value is detected.

It must be noted that these two families of control charts, the Shewhart charts on the one hand and the CUSUM and EWMA charts on the other hand, are complementary.

Indeed, the Shewhart charts aim to detect only large shifts above 1.5 $s(x_i)$ (detection of outliers), whereas the CUSUM and EWMA charts aim to detect only small shifts below 1.5 $s(x_i)$ and more generally drifts in the process. In other words, Shewhart charts perform an analysis in the high-frequency domain, whereas the CUSUM and EWMA charts are more suitable to analyze in the low-frequency domain.

3. Test of control charts with simulated data

3.1 The Gaussian white noise model

$N = 100$ normally distributed data points x_i are simulated using the random generator from the numpy.py library.

The model of the simulated process is the following:

$$\begin{cases} x_i = \mu_i + \varepsilon_i \\ \varepsilon_i \sim \mathcal{N}\left(0, \sigma_i^2\right) \end{cases} \tag{9}$$

with μ_i: the expectation parameter at the time t_i, $\mathcal{N}$: the normal distribution, and σ_i^2: the variance parameter of the measurement noise ε_i at the time t_i. It is referred below as a Gaussian white noise model.

The reference dataset is defined by setting, for all $i \in [\![1, N]\!]$, the parameters $\mu_i = 0$ and $\sigma_i^2 = 1$. The control charts are presented below, and the tests conclude—as expected—that the simulated process is under statistical control (**Figures 1–4**).

3.2 The Gaussian white noise model with a deterministic drift

Now, a slight positive drift is considered in the following model referred to as a deterministic drift model in a Gaussian white noise such as the expectation parameter takes values such as:

$$\begin{cases} \mu_i = 0, & \forall i \in \left[\!\left[1, \frac{N}{2}\right]\!\right] \\ \mu_i = \mu_{i-1} + 0.3, & \forall i \in \left[\!\left[\frac{N}{2} + 1, N\right]\!\right]. \end{cases} \tag{10}$$

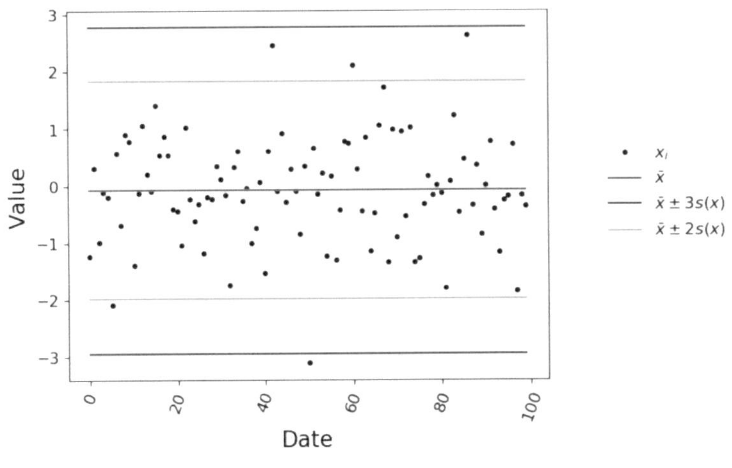

Figure 1.
x-chart of the process simulated form the Gaussian white noise model.

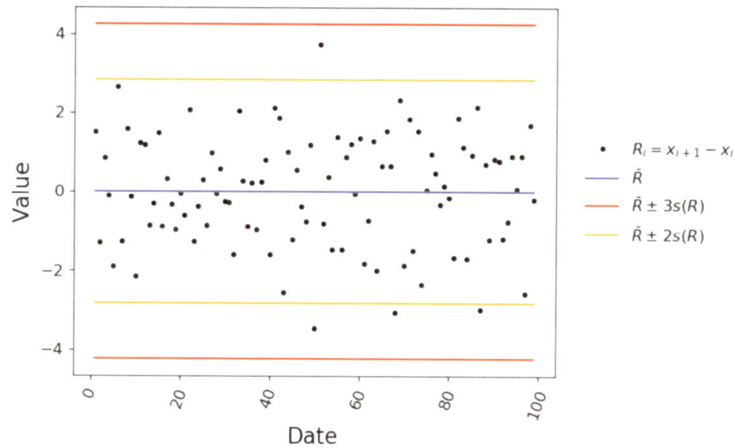

Figure 2.
R-chart of the process simulated form the Gaussian white noise model.

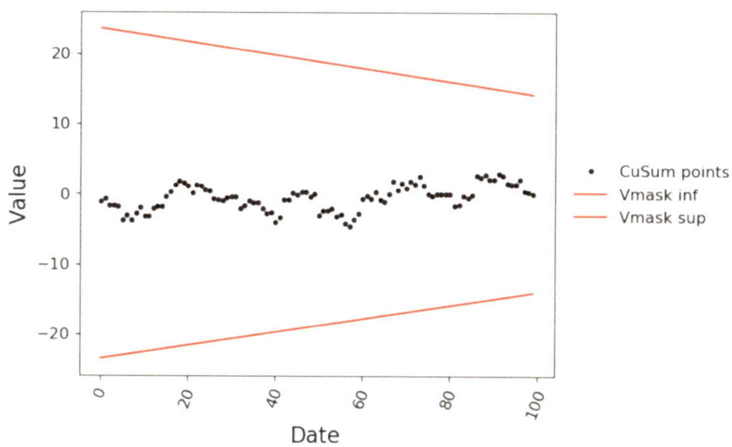

Figure 3.
CUSUM chart of the process simulated form the Gaussian white noise model.

This leads to a detection of this trend in the x-chart because more than nine points are in a row on one side of the central line. Also, the CUSUM chart and the EWMA chart detect the trend with some points out of their control limits (**Figures 5–8**).

3.3 The autocorrelated model

The third model studied here is the autocorrelated model (or the colored noise model) where a low-frequency Brownian noise is introduced in the following autocorrelated model:

$$\begin{cases} x_i = x_{i-1} + \dot{x}_i + \varepsilon_i \\ \dot{x}_i = \dot{x}_{i-1} + \delta_i \end{cases} \tag{11}$$

where the measurement noise is $\varepsilon_i \sim \mathcal{N}\left(0, \sigma_i^2\right)$ (idem as in Eq. (9)) and a process noise $\delta_i \sim \mathcal{N}\left(0, \tau_i^2\right)$ is added with a variance parameter τ_i^2. In the example below, $\tau_i = 0.3$ while σ_i is still equal to 1.

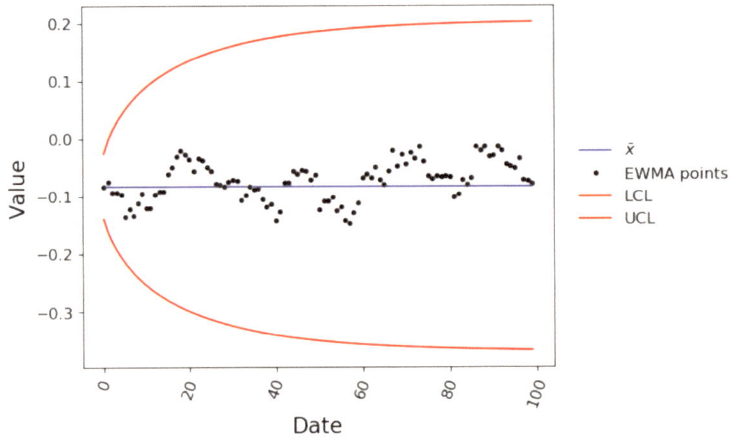

Figure 4.
EWMA chart of the process simulated form the Gaussian white noise model.

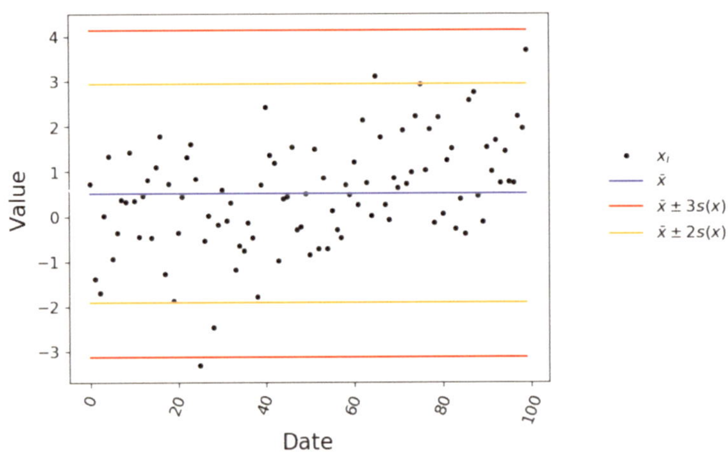

Figure 5.
x-chart of the process simulated form the Gaussian white noise model with a slight drift.

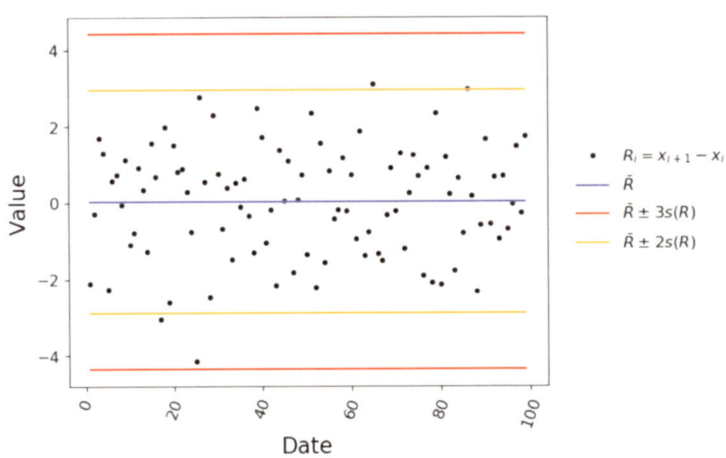

Figure 6.
R-chart of the process simulated form the Gaussian white noise model with a slight drift.

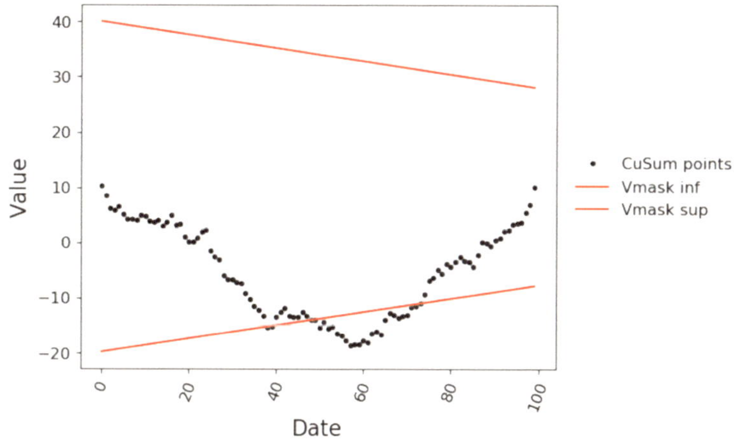

Figure 7.
CUSUM chart of the process simulated form the Gaussian white noise model with a slight drift.

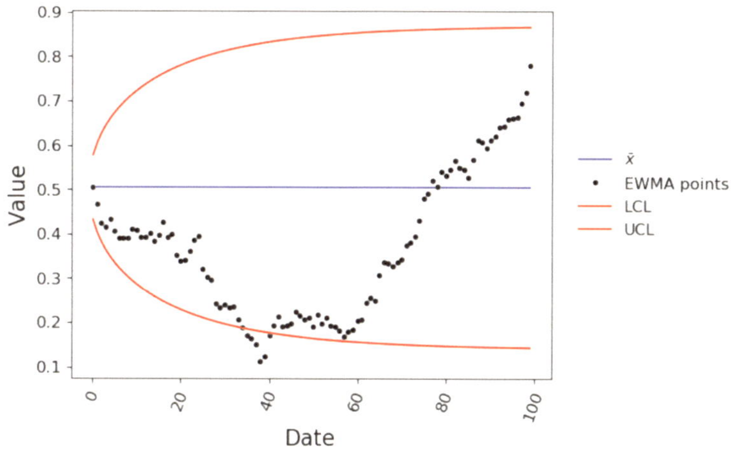

Figure 8.
EWMA chart of the process simulated form the Gaussian white noise model with a slight drift.

Again, this type of deviation from the Gaussian white noise model is detected by the value chart with more than nine points in a row on one side of the central line and by CUSUM and EWMA charts (**Figures 9–12**).

These three examples underline the fact that control charts are—by definition—build to verify whether a dataset fits with a Gaussian white noise model or not. Any deviation from this model is efficiently revealed by control charts. An operator could therefore conclude by the fact that the process is not under statistical control. This conclusion is good in the case of deterministic trends that are that a metrology system seeks to avoid maintaining the references that it produces. However, it has been demonstrated that the autocorrelated noise could also lead to an alert by control charts. On the contrary to the deterministic drift, the latter is a natural random process that appears in every system, which is observed on enough long-term periods. This particular limit of control charts was already pointed out by many authors [7–10]. This issue is an emphasis in the real cases presented below.

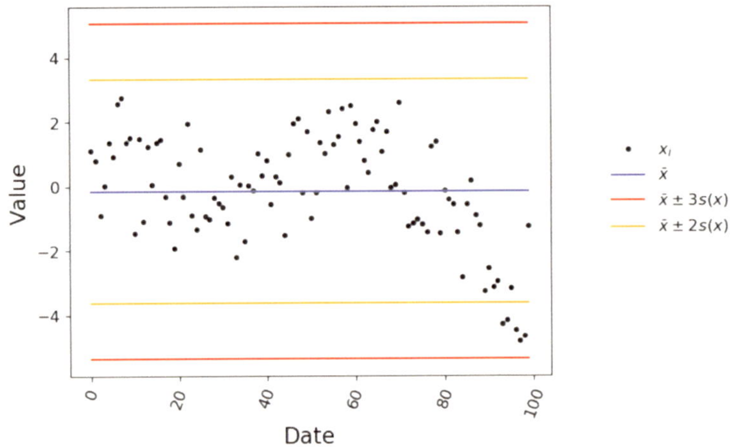

Figure 9.
x-chart of the process simulated form the autocorrelated model.

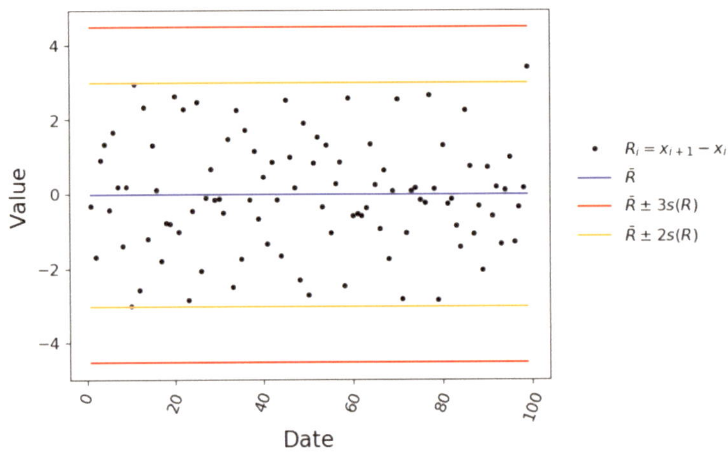

Figure 10.
Range chart of the process simulated form the autocorrelated model.

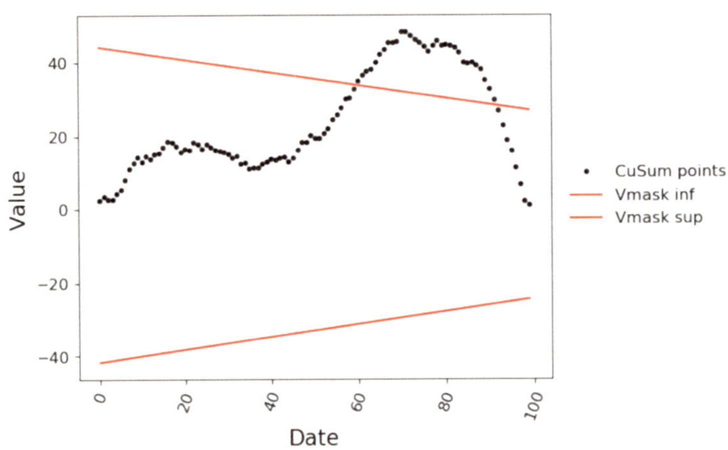

Figure 11.
CUSUM chart of the process simulated form the autocorrelated model.

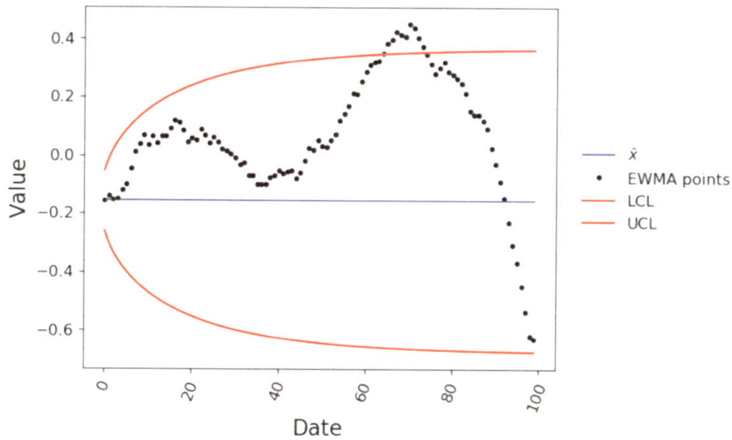

Figure 12.
EWMA chart of the process simulated form the autocorrelated model.

4. Test of control charts with real data

Two data are here considered. The first one comes from the periodic control of a recent metrology system that operates at the BIPM for radionuclide metrology and is called the ESIR (extended SIR) [11, 12]. The second example comes from the periodic control of a metrology service, the SIR, which operates at the BIPM for radionuclide metrology since 1976 [13].

4.1 Test with 1-year period data

The ESIR is under development, and the periodical control using check standards has been set up for 1 year. **Figures 13–16** show the analysis of the measured data by control charts. The monitored value is a comparison indicator obtained by using a check source of ^{14}C. The analysis from the four-control charts leads to the conclusion that the process is under statistical control. In other words, the measurement model complies with a Gaussian white noise model. At least, it is true for this short period of observation. The activity estimation is of the ^{14}C check source is stable and equal to 24.3373 ± 0.0184 A.U. There is no unit associated with the

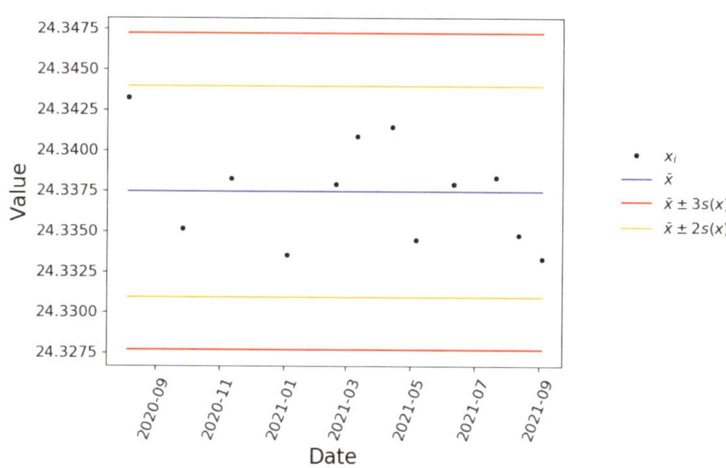

Figure 13.
x-chart of the periodic control of the ESIR metrology system.

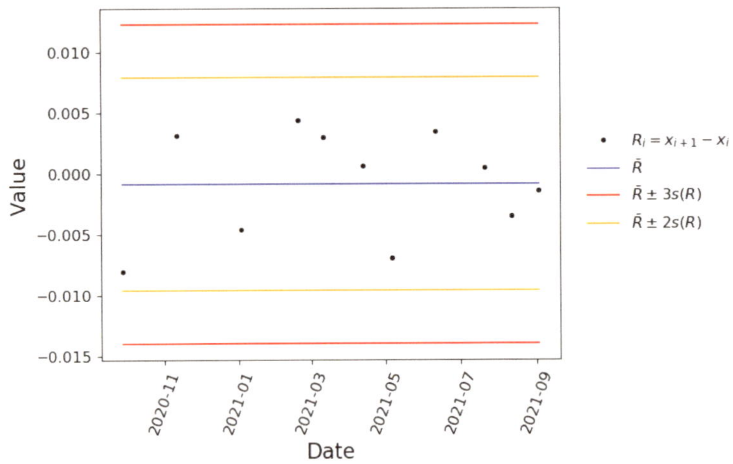

Figure 14.
R-chart of the periodic control of the ESIR metrology system.

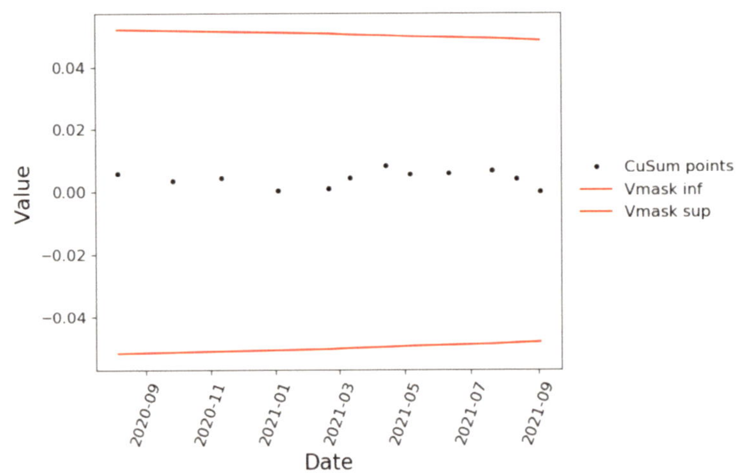

Figure 15.
CUSUM chart of the periodic control of the ESIR metrology system.

measurand because the aim of the ESIR is only to deliver a measure proportional to the activity but is very reproducible during the time. The relative standard uncertainty associated with this measurement is about 0.08%.

4.2 Test with 20-year period data

The SIR produces international references for decades. The ^{226}Ra source is used as a control standard to monitor the reproducibility of the measured current. In this case, the control charts are applied to a 20-year period of data acquisition.
They have detected that the process is not under statistical control. However, these low-frequency fluctuations could certainly be a random colored noise and not a deterministic drift. This phenomenon appears when measurement values are observed over a very long period such as here. The only conclusion that can be drawn from control charts is that the data do not fit with the Gaussian white noise model and that a type A evaluation of the uncertainty associated with the measurement values (from t_1 to t_N) cannot be applied. Nevertheless, it is no need to refine

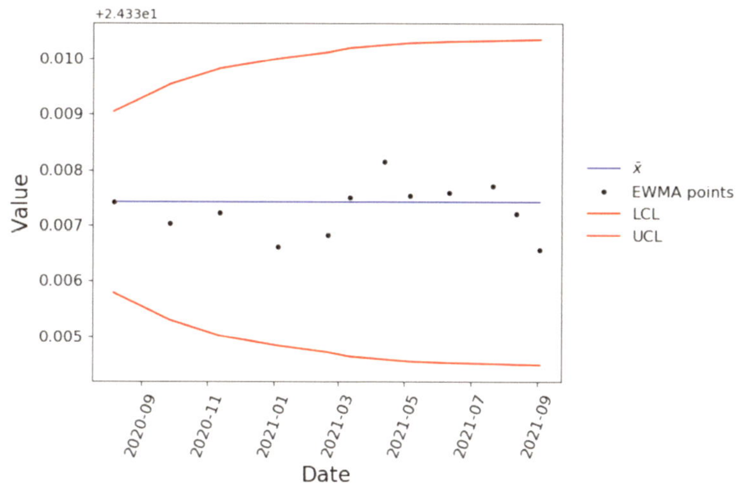

Figure 16.
EWMA chart of the periodic control of the ESIR metrology system.

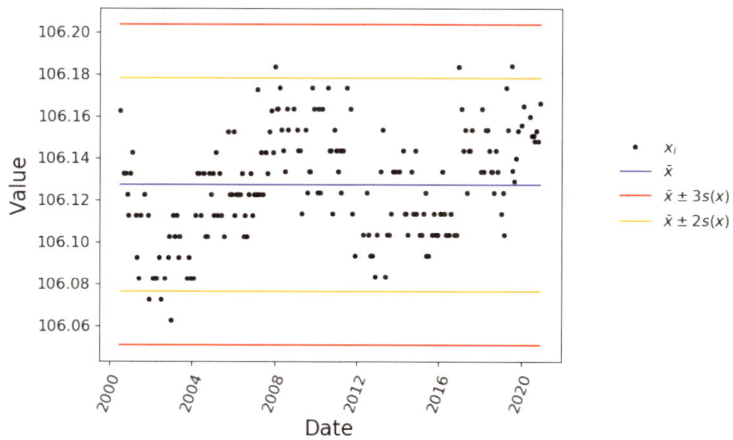

Figure 17.
x-chart of the periodic control of the SIR metrology system.

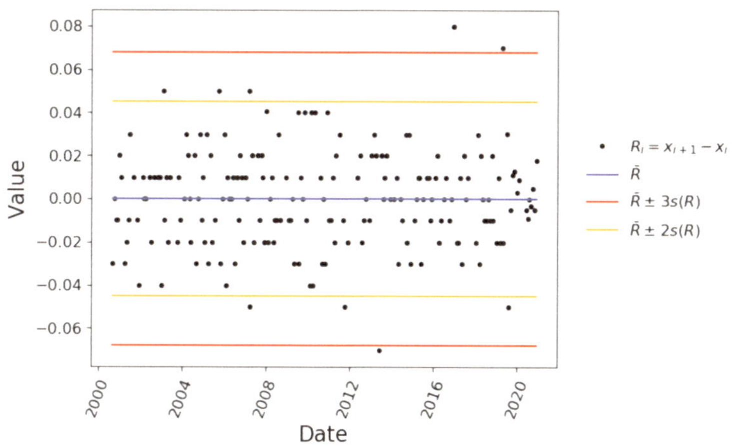

Figure 18.
R-chart of the periodic control of the SIR metrology system.

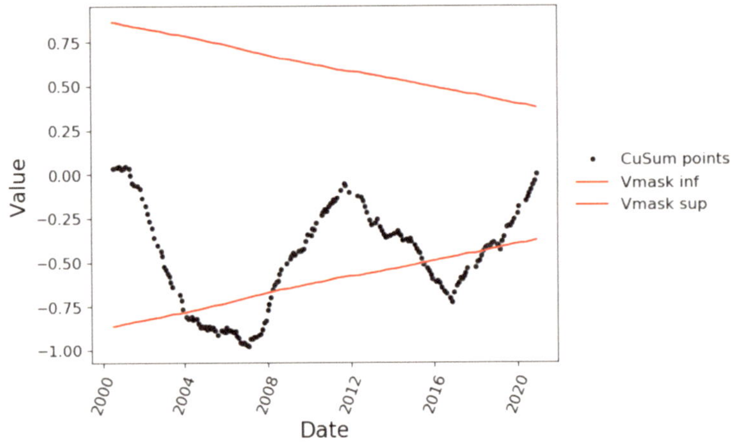

Figure 19.
CUSUM chart of the periodic control of the SIR metrology system.

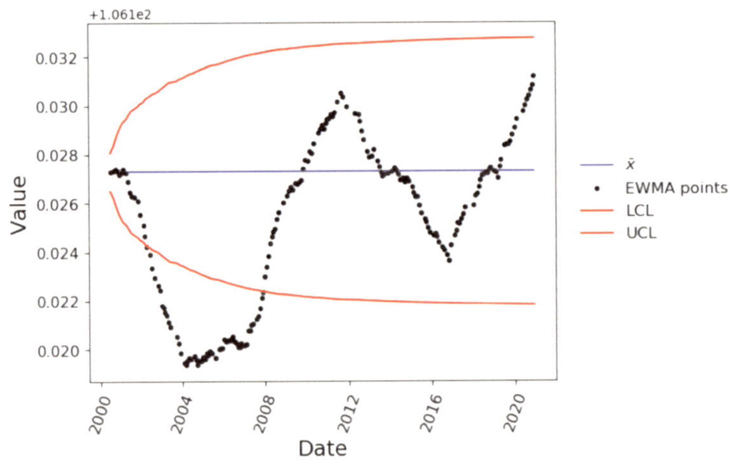

Figure 20.
EWMA chart of the periodic control of the SIR metrology system.

the measurement model for the SIR measurement because the quantity delivered by the system is a ratio of a punctual current measurement (at a time t_i) from which this low-frequency component revealed here is suppressed by the method (**Figure 17–20**).

We have underlined, through these simulated and real datasets, the limitation of control charts to deal with the evaluation of the long-term stability of metrology instruments. In this case, time series analysis could be used as more appropriate tool to address this objective.

An example of time series analysis is introduced below.

5. Introduction to time series analysis applied to a long-term measurement model

It has been seen that the autocorrelated model introduced in Eq. (10) is appropriate to represent measurement data when observed over a long-term period.

Eq. (10) is here rewritten in a state-space model where the measurement is decomposed by the following:

- the fluctuation of the expectation value $(x_1, \dots, x_i, \dots, x_N)$ with the low-frequency random noise at measurement times $(t_1, \dots, t_i, \dots, t_N)$,

- and the measurements gathered in the vector $(y_1, \dots, y_i, \dots, y_N)$ where the Gaussian white noise is finally added to $(x_1, \dots, x_i, \dots, x_N)$.

The model becomes,

$$\begin{cases} x_i = x_{i-1} + \dot{x}_{i-1} \\ \dot{x}_i = \dot{x}_{i-1} + \delta_i \\ y_i = x_i + \varepsilon_i \end{cases} \tag{12}$$

The drift velocity $\dot{x}_i = x_i - x_{i-1}$ is introduced such that the dynamic Gaussian noise $\delta_i \sim \mathcal{N}(0, \tau^2)$ describes a random walk parametrized by a variance τ^2. And the measurement noise $\varepsilon_i \sim \mathcal{N}(0, \sigma^2)$ is Gaussian white noise with a variance σ^2.

The idea behind time series analysis is to deconvolve the low-frequency fluctuation of the expectation values $(x_1, \dots, x_i, \dots, x_N)$ from the observations $(y_1, \dots, y_i, \dots, y_N)$.

At first, the measurement variance σ^2 is evaluated by using the median of all the punctual type A evaluation of the standard deviation $s(y_i)$ which are (by definition) isolated from the dynamic component.

$$\sigma^2 = \text{median}\left(s^2(y_i)\right) \tag{13}$$

with,

$$s^2(y_i) = \frac{\sum_{j=1}^{M}\left(y_i^{(j)} - M^{-1}\sum_{l=1}^{M} y_i^{(l)}\right)^2}{M-1}. \tag{14}$$

$\left(y_{N+1}^{(1)}, \dots, y_{N+1}^{(j)}, \dots, y_{N+1}^{(M)}\right)$ is the set of measurements obtained at the time t_i during a period that allowing type A evaluation according to the Gaussian white noise model. Here, control charts can be used to verify this hypothesis.

Eq. (12) is transformed in the following matrix form:

$$\begin{cases} X_i = FX_{i-1} + W_i \\ y_i = HX_i + \varepsilon_i \end{cases} \tag{15}$$

with the "state vector" $X_i = \begin{pmatrix} x_i \\ \dot{x}_i \end{pmatrix}$, the "transition matrix" $F = \begin{pmatrix} 1 & 1 \\ 0 & 1 \end{pmatrix}$, the" dynamic noise matrix" $W_k = \begin{pmatrix} 0 \\ \delta_i \end{pmatrix}$, and the "observation matrix" $H = \begin{pmatrix} 1 & 0 \end{pmatrix}$.

5.1 The Kalman filter

Given the Gaussianity of the noises (both dynamic noise δ_i and measurement noise ε_i) of this linear model (in the sense of time series), the Kalman filter (KF) provides an optimal estimation of the states X_i given the current and past measurements $y_{1:i}$. The optimality of KF to solve Eq. (15) is established through both least-square interpretation and the probabilistic Bayesian analysis [14, 15]. Hence, KF is a recursive Bayesian filter where the states X_i and the measurements y_i are Markov processes implying

$$\begin{cases} p(X_i|X_{1:i-1}) = p(X_i|X_{i-1}) \\ p(y_i|X_{1:i}) = p(y_i|X_i). \end{cases} \tag{16}$$

KF evaluates, at each time step i, the parameters of the Gaussian posterior distribution $p(X_i|y_{1:i})$ in two steps:

- The first step consists of predicting the state X_i given past measurement data $y_{1:i-1}$. The posterior distribution $p(X_{i-1}|y_{1:i-1})$ at t_{i-1} is used to build a prior distribution at the time t_i according to the linear model (see Eq. (15)) from t_{i-1} to t_i such as

$$p(X_i|y_{i-1}) = \int p(X_i|X_{i-1})\, p(X_{i-1}|y_{i-1})\, \mathrm{d}X_{i-1}. \tag{17}$$

- The second step aims to update the posterior distribution at t_i given the predicted prior $p(X_i|y_{1:i-1})$, and the likelihood $p(y_i|X_i)$ of the measurement y_i given the estimate X_i.

$$p(X_i|y_i) \propto p(X_i|y_{i-1})\, p(y_i|X_i). \tag{18}$$

The posterior state estimate $\hat{X}_{i|i}$ and the posterior of the covariance $P_{i|i}$ of the posterior distribution $p(X_i|y_{1:i})$ are evaluated through the Kalman filtering process described below.

From $i = 1$ to $i = N$, KF predicts the state X_i and the measurement y_i at the time step i given the previous state at $i - 1$:

$$\begin{cases} \hat{X}_{i|i-1} = F\hat{X}_{i-1|i-1} \\ \hat{y}_{i|i-1} = H\hat{X}_{i|i-1}. \end{cases} \tag{19}$$

With the state and measurement covariances, P_i and S_i such that:

$$\begin{cases} P_{i|i-1} = FP_{i-1|i-1}F^{\mathrm{T}} + \tau^2 \\ S_i = HP_{i|i-1}H^{\mathrm{T}} + \sigma^2. \end{cases} \tag{20}$$

Then, the pre-fit residual (innovation) e_i of the estimation and the Kalman gain K_i are calculated by

$$\begin{cases} e_i = y_i - \hat{y}_{i|i-1} \\ K_i = P_{i|i-1}H^{\mathrm{T}}S_i^{-1}. \end{cases} \tag{21}$$

Finally, the updated state estimation is

$$\begin{cases} \hat{X}_{i|i} = \hat{X}_{i|i-1} + K_i e_i \\ P_{i|i} = (I - K_i H)P_{i|i-1}, \end{cases} \tag{22}$$

with I: the identity matrix.

5.2 The Rauch-Tung-Striebel smoother

The KF estimates the states $\hat{X}_{i|1:i}$ given present and past observations from t_1 to t_i. To refine the estimation, a KF backward pass (smoothing pass) processes the filtered estimates data from t_N to t_1. This procedure is called the Rauch-Tung-Striebel smoother (RST) [16, 17]. Symmetrically to the forward KF optimal filter, the optimal smoothing also consists in:

- a prediction step where the state X_{i+1} is estimated given measurement data $y_{1:i}$ and the forecasted state backward transition according to the linear model from t_{i+1} to t_i.

$$p(X_{i+1}|y_{1:i}) = \int p(X_{i+1}|X_i)p(X_i|y_{1:i})\mathrm{d}X_i \tag{23}$$

- The second step aims to update the posterior distribution at t_i given all the data $p(X_i|y_{1:N})$.

$$p(X_i|y_{1:N}) = p(X_i|y_{1:i})\int \frac{p(X_{i+1}|X_i)p(X_{i+1}|y_{1:N})}{p(X_{i+1}|y_{1:i})}\mathrm{d}X_{i+1}. \tag{24}$$

The RTS smoother calculates the Maximum A Posteriori (MAP) estimate $\hat{X}_{i|1:N}$ of the posterior probability $p(X_i|y_{1:N})$ given all the observations (past, present, and future) from t_1 to t_N.

From $i = N$ to $i = 1$, the residual e'_i between the filtered state $\hat{x}_{i+1|i}$ and the prediction $\hat{x}_{i+1|1:N}$ is used to calculate a new Kalman gain G_i such as

$$\begin{cases} e'_i = \hat{X}_{i+1|1:N} - \hat{X}_{i+1|i} \\ G_i = P_{i|i+1}F^{\mathrm{T}}P^{-1}_{i+1|i}. \end{cases} \tag{25}$$

The optimal state estimation is finally given by

$$\begin{cases} \hat{X}_{i|1:N} = \hat{X}_{i|i} - G_i e'_i \\ P_{i|1:N} = P_{i|i} + G_i(P_{i+1|1:N} - P_{i+1|i})G_i^{\mathrm{T}}. \end{cases} \tag{26}$$

5.3 The auto-tuning of the RTS smoother

As the measurement noise variance σ^2 is evaluated, by definition, through measurement repeated during a period free from long-term fluctuation (Eq. (17)), the variance of the dynamic noise τ^2 is the remaining unknown parameter of the model. The auto-tuning procedure proposed here to evaluate τ starts with a loss function L defined by

$$L(\tau) = \sum_{i=2}^{N} e_i^2 + \left(\hat{x}_{i|1:N} - \hat{x}_{i-1|1:N}\right)^2. \tag{27}$$

The value $\hat{\tau}$ minimizing $L(\tau)$ aims to minimize the residual between the measurements y_i and the estimations of the measurements $\hat{y}_{i|i-1}$ as well as the distance between two consecutive filtered estimations $\hat{x}_{i|1:N}$ and $\hat{x}_{i-1|1:N}$. On the one hand, if this last smoothing constraint is not considered, τ will diverge toward high values describing a totally unstable system. On the other hand, if the prediction error of the model is not considered, τ will be canceled by describing a stable system with constant $\hat{x}_{i|i}$ values. So, the loss function of Eq. (27) permits to produce a model of measurement of a controlled process containing a low-frequency dynamic noise. The variance of the low-frequency dynamic noise is determined through an iterative procedure minimizing $L(\tau)$ such that,

$$\hat{\tau} = \underset{\tau}{\text{argmin}}(L(\tau)). \tag{28}$$

The auto-tune RTS smoother provides filtered values $\hat{x}_{i|1:N}$ where the measurement noise, ε_i, is removed given the knowledge of the variances of the measurement noise σ^2 (see Eqs. (13) and (14)) and the dynamic noise $\hat{\tau}^2$. So, the global hidden uncertainty component from the long-term process random noise can be evaluated from the deconvolved data $\hat{x}_{i|1:N}$ such that,

$$s\left(\hat{x}_{i|1:N}\right) = (N-1)^{-1}\sum_{i=1}^{N}\left(\hat{x}_{i|1:N} - N^{-1}\sum_{j=1}^{N}\hat{x}_{i|1:N}\right)^2. \tag{29}$$

Thanks to the intrinsic unfolding of the measurement noise (Gaussian white) and the process noise (Gaussian red) defined in the dynamic model (Eq. 12), the standard uncertainty $u(x_{N+1})$ of a new measurement at t_{N+1} can be obtained by the quadratic summation of the type A uncertainty evaluation $s\left(x_{i(N+1)}\right)$ obtained under repeated conditions $\left(y_{N+1}^{(1)}, \dots, y_{N+1}^{(j)}, \dots, y_{N+1}^{(M)}\right)$ and the type B uncertainty evaluation evaluated using all historical data $(y_1, \dots, y_i, \dots, y_N)$ through the above-described auto-tunned RTS smoother.

Therefore, the standard uncertainty $u^2(x_{N+1})$ of the next measurement at t_{N+1} should be inflated considering the impact of the dynamic noise such that,

$$u^2(x_{T+1}) = s^2\left(x_{i(T+1)}\right) + P_{N+1|1:N}(1) \tag{30}$$

With,

$$P_{N+1|1:N} = FP_{N|1:N}F^{\mathrm{T}} + \tau^2 \tag{31}$$

This auto-tuning of a RTS smoother applied to measurement data recorded over a long-term period fully address the recommendation of the JCGM100:2008 guide [4]:

"Because the mathematical model may be incomplete, all relevant quantities should be varied to the fullest practicable extent so that the evaluation of uncertainty can be based as much as possible on observed data. Whenever feasible, the use of empirical models of the measurement founded on long-term quantitative data, and the use of check standards and control charts that can indicate if a measurement is under statistical control, should be part of the effort to obtain reliable evaluations of uncertainty. The mathematical model should always be revised when the observed data, including the result of independent determinations of the same measurand, demonstrate that the model is incomplete. A well-designed experiment can greatly facilitate reliable evaluations of uncertainty and is an important part of the art of measurement."

6. Application of time series analysis on simulated and real measurement data

The time series analysis through the auto-tuned RST smother is presented in **Figure 21** for the simulated dataset from Section 3.3 obtained with the

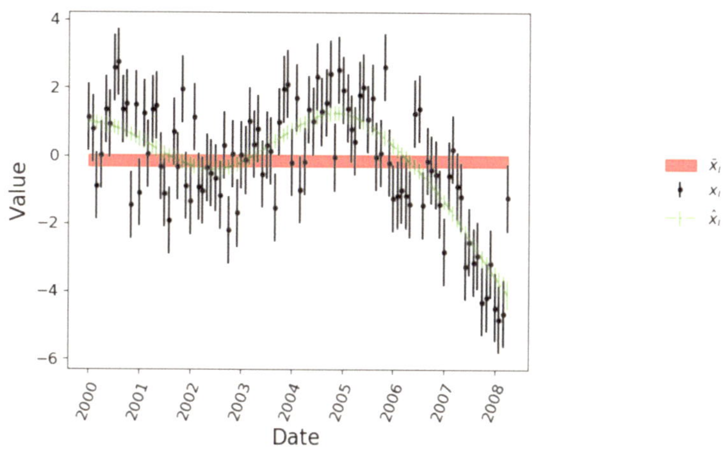

Figure 21.
Time series analysis of the measurement data simulated form the autocorrelated model.

autocorrelated model. Black points are the measurement points y_i with error bars representing the standard deviation $s(y_i)$. Green points are the estimations of the low-frequency fluctuations (red noise) of the measurement $\hat{x}_i$ with error bars representing the standard deviation $s(\hat{x}_i)$. The red line is the global expectation value where the thickness of the line represents +/− 1 standard deviation.

It can be seen the good deconvolution of the low-frequency random noise shown in green points. This leads to a standard uncertainty from this autocorrelated noise, $s(\hat{x}_{i|1:N}) = 1.31$. This value is significant when compared to type A evaluation of individual measurement points $\sigma^2 = 1.74$.

Also, the time series analysis through the auto-tuned RST smother is presented in **Figure 22** when applied to the periodic control of the SIR metrology system. Again, black points are the measurement points y_i with error bars representing the standard deviation $s(y_i)$. Green points are the estimations of the low-frequency fluctuations (red noise) of the measurement $\hat{x}_i$ with error bars representing the standard deviation $s(\hat{x}_i)$. The red line is the global expectation value where the thickness of the line represents +/− 1 standard deviation.

The analysis using control charts (see Section 4.2) has led to conclude that the system is not under statistical control although autocorrelated noise is expected for

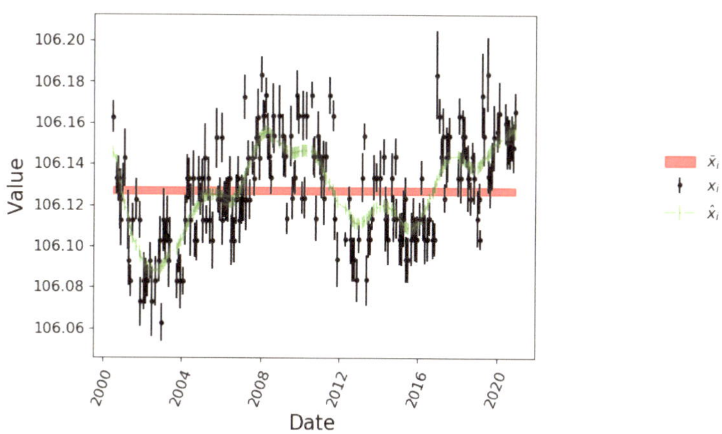

Figure 22.
Time series analysis of the periodic control of the SIR metrology system.

such a long-term period of monitoring. The optimal Bayesian smoother permits to properly deconvolve this component (see green points) and to evaluate the standard uncertainty associated with it: $s(\hat{x}_{i|1:N}) = 0.0183$. This value is significant when compared to type A evaluation of individual measurement points $\sigma^2 = 0.025$.

7. Conclusion

It has been emphasized through simulated and experimental data that control charts (Shewhart, CUSUM, EWMA charts) are convenient tools to assess whether a measurement system is under statistical control or not. They test the compliance of the measured dataset with a Gaussian white noise model. It can notably be used to determine the possibility to apply a type A evaluation of uncertainty on the measurement data.

However, care should be taken when monitoring over long periods of time. A low-frequency random noise will inevitably appear. Despite its stochastic nature, the latter will induce detection by control charts. In this case, it is not appropriate to conclude by a loss of control of the system but rather to re-evaluate the measurement model to include this long-term component. To that end, time series analysis could be implemented to better deal with this particular case of measurement data.

An auto-tuned Rauch-Tung-Striebel smoother, based on the optimal Bayesian filter called Kalman filter, is introduced to illustrate how time series analysis can help to deconvolve the components of the long-term measurement model. The low-frequency random fluctuation (aka red noise) can be estimated and used to evaluate the standard uncertainty component it induces on the measurement.

Author details

Romain Coulon
Bureau International des Poids et Mesures, Pavillon de Breteuil, Sèvres Cedex, France

*Address all correspondence to: romain.coulon@bipm.org

References

[1] Milton MJT, Possolo A. Trustworthy data underpin reproducible research. Nature Physics. 2020;**6**:117-119

[2] ISO/IEC 17025:2017 General requirements for the competence of testing and calibration laboratories

[3] Wikipedia Seven basic tools of quality

[4] JCGM-BIPM. Evaluation of Measurement Data—Guide to the Expression of Uncertainty in Measurement. Vol. 50. Geneva: International Organization for Standardization; 2008. p. 134

[5] Anon ISO 7870-1:2019(en) Control charts—Part 1: General guidelines

[6] Prins JC, Croarkin PT. What are Control Charts? Methods, NIST/SEMATECH E-handb. Stat 6.3.1. Gaithersburg: NIST; 2012

[7] Leitao S, Emilia M, Pedro J. Applications of control charts ARIMA for autocorrelated data. In: Practical Concepts of Quality Control. IntechOpen; 2012

[8] Bagshaw M, Johnson RA. The effect of serial correlation on the performance of CUSUM tests II. Technometrics. 1975; **17**:73

[9] Tamhane AC, Malthouse EC. Control charts for autocorrelated process data. In: Advances in Statistical Decision Theory and Applications. Boston, MA: Birkhäuser Boston; 1997. pp. 371–385

[10] Umar Farouk A, Mohamad I. The enhanced EWMA control chart with autocorrelation. Jurnal Teknologi. 2014;**71**:89-94

[11] Coulon R, Broda R, Cassette P, Courte S, Jerome S, Judge S, et al. The international reference system for pure β-particle emitting radionuclides: An investigation of the reproducibility of the results. Metrologia. 2020;**57**: 035009

[12] Coulon R, Judge S, Liu H, Michotte C. The international reference system for pure beta-particle emitting radionuclides: an evaluation of the measurement uncertainties. Metrologia. 2021;**58**:025007

[13] Ratel G. The système international de référence and its application in key comparisons. Metrologia. 2007;**44**:S7-S16

[14] Jazwinski A. Stochastic Processes and Filtering Theory (Pittsburgh). New York and London: Academic Press; 1970

[15] Bar-Shalom Y, Li X-R, Kirubarajan T. Estimation with Applications to Tracking and Navigation. New York, USA: John Wiley & Sons, Inc.; 2001

[16] Rauch H. Solutions to the linear smoothing problem. IEEE Transactions on Automatic Control. 1963;**8**:371–372

[17] Rauch HE, Tung F, Striebel CT. Maximum likelihood estimates of linear dynamic systems. AIAA Journal. 1965;**3**: 1445–1450

Chapter 2

The Data Evaluation of Interlaboratory Comparisons for Calibration Laboratories

Oleh Velychko and Tetyana Gordiyenko

Abstract

National accreditation agencies in different countries have set quite strict requirements for accreditation of testing and calibration laboratories. Interlaboratory comparisons (ILCs) are a form of experimental verification of laboratory activities to determine technical competence in a particular activity. Successful results of conducting ILCs for the laboratory are a confirmation of competence in carrying out certain types of measurements by a specific specialist on specific equipment. To obtain reliable results of ILC accredited laboratories, it is necessary to improve the methods of processing these results. These methods are based on various data processing algorithms. Therefore, it is necessary to choose the most optimal method of processing the obtained data, which would allow to obtain reliable results. In addition, it is necessary to take into account the peculiarities of the calibration laboratories (CLs) when evaluating the results of ILC. Such features are related to the need to provide calibration of measuring instruments for testing laboratories. The evaluation results for ILCs for CLs are presented. The results for all participants of ILCs were evaluated using the E_n and z indexes. The obtained results showed that for the such ILCs it is also necessary to evaluate the data using the z index also.

Keywords: interlaboratory comparison, data evaluation, referent laboratory, calibration laboratory, calibration, measurement uncertainty

1. Introduction

Participants in the International Laboratory Accreditation Cooperation (ILAC) Mutual Recognition Agreement (MRA) recognize the calibration or test results obtained by each other's accredited calibration and testing laboratories [1–4]. ILAC Policy and Procedural publications are for the operation of the ILAC MRA. ILAC has a special policy for participation in proficiency testing activities, on metrological traceability of measurement results, for measurement uncertainty in calibration [5–7]. The policy for measurement uncertainty to base on the Guide to Uncertainty in Measurement (GUM) [8–11] and retains the common understanding of the term calibration and measurement capabilities (CMCs) from the joint declaration issued by the International Bureau of Weights and Measures (BIPM) and ILAC [12]. ILAC has a special guideline for measurement uncertainty in testing [13]. This document provides guidance for the evaluation and reporting of measurement uncertainty in testing accordance with the requirements of the International Standard ISO/IEC 17025 [14].

National accreditation agencies in different countries have set quite strict requirements for accreditation of testing and calibration laboratories. Laboratory accreditation criteria in most accreditation systems include three main groups: laboratory technical equipment, personnel competence, and the effectiveness of the quality system. Interlaboratory comparisons (ILCs) are a form of experimental verification of laboratory activities to determine technical competence in a particular activity. Successful results of conducting ILCs for the laboratory are a confirmation of competence in carrying out certain types of measurements by a specific specialist on specific equipment.

To obtain reliable results of ILC accredited laboratories, it is necessary to improve the methods of processing these results. These methods are based on various data processing algorithms as required by international and regional guidelines and standards. To conduct ILC for CLs, it is necessary to take into account the relevant requirements of the international standards ISO/IEC 17025 [14] and ISO/IEC 17043 [15]. Therefore, it is necessary to choose the most optimal method of processing the obtained data, which would have a minimum number of restrictions on the application and allow to obtain reliable results. In addition, it is necessary to take into account the peculiarities of the calibration laboratories (CLs) when evaluating the results of ILC. Such features are related to the need to provide calibration of measuring instruments for testing laboratories.

ILCs for CLs are held nationally in different countries. Such ILCs are carried out to establish the competence of the CLs in calibrating various measuring instruments and working standards for various measured quantities [16–24]. For their implementation, various calibration objects are used. To evaluate the ILC data, various methods of their data processing are used [25–30], and to estimate the measurement uncertainty, the regional guidance EA-04/02 M [31] is additionally used, in addition to the ILAC documents [8, 13]. However, in addition to the method of data evaluation, it is necessary to take into account other influencing factors on the CL result of ILC. In particular, unsatisfactory ILC results for all participating CLs may be associated with a large time drift of the calibrated measuring instrument.

The growing practical need of ILCs for CLs to ensure recognition of the obtained results at both national and international levels underscores the relevance of this research.

2. The national interlaboratory comparisons for calibration laboratories

The main purpose of accredited CLs is to calibrate working standards and measuring instruments for accredited testing laboratories. Significantly more testing laboratories are accredited by national accreditation bodies than CLs. For example, at the middle of 2021, 837 testing and 35 calibration laboratories were accredited in Ukraine. This represents only 4% of accredited CLs of the total number of all accredited laboratories. Therefore, the number of ILCs for testing laboratories is objectively much larger than for CLs.

The State Enterprise "Ukrmetrteststandard" (Ukraine) as a referent laboratory (RL) organized and carry out seven ILCs for accredited CLs from 2016 to 2019 [32–35, etc]. The list of these ILCs is shown in **Table 1**. The calibration objects for these ILCs were working standards and measuring instruments for electrical quantities, and time and frequency. When carrying out comparisons, CLs calibrated objects in accordance with the requirements of the international standard ISO/IEC 17025 [14]. The total number of calibration object parameters ranged from 3 to 12. The total number of CLs with RL that took part in these comparisons ranged from 5 to 10.

ILC	Calibration object	Number of parameters	Number of participants	Period of carrying out
ILC1	Precision measuring thermocouple	AC voltage at 5 frequencies	5 labs	2016–2018
ILC2	Measures of electrical resistance (1th round)	3 nominations of resistance	8 labs	2016
ILC3	Measures of electrical resistance (2th round)	3 nominations of resistance	5 labs	2018–2019
ILC4	Precision measure of electric power	6 power factors at 2 frequencies	8 labs	2016–2018
ILC5	Low frequency signal generator	AC voltage at one frequency, total harmonic factor at 4 frequencies, 5 frequencies	4 labs	2016
ILC6	Electronic stopwatch	3 time intervals	9 labs	2016
ILC7	High-frequency signal generator	3 frequencies	10 labs	2018

Table 1.
The list of national ILCs for CLs.

In all presented ILCs, the assigned value (AV) with its uncertainty was taken as the value with its uncertainty of the RL. This was done because the RL had the best measurement capabilities among all CLs that took part in the comparisons. For many years RL has taken part in international comparisons of national measurement standards of electrical quantities within the framework of Regional Metrological Organizations (COOMET, EURAMET, and GULFMET) and had positive results. RL also had published CMCs for some electrical quantities in the BIPM Key Comparison Database [36].

A program for all ILCs was implemented in accordance with the requirements of ISO/IEC 17043 [15]. CLs that participated in the ILCs performed calibration of the measuring instruments (calibration object) provided to the RL in accordance with their own methods according to the radial scheme [4]. RL sent the calibration object to the participating laboratory, and this laboratory returned this object back to RL. In this case, the RL constantly monitored the stability of the calibration object [35, 37]. The RL determined the characteristics of the instability of the calibration object before and after its research in the CLs participating in the ILC.

In accordance with the adopted ILC programs, RL analyzed the calibration data provided by the CLs [38], in particular, analyzed the declared measurement uncertainty. The data obtained from CLs were necessarily checked by RL for their consistency. Indicators for assessing of consistency were E_n and z indexes set and defined in [4, 15]. The general algorithm of processing of the received primary data of ILCs given in [34] was used. In case of inconsistent data, RL reported this appropriate CL and analyzed the responses received from this laboratory. RL prepared a report on the comparisons, evaluating the data of all CLs. In the event that the laboratory or laboratories received inconsistent results of comparisons, RL suggested that they take the necessary corrective action.

3. The traditional data evaluation of interlaboratory comparisons

The traditional assessment of ILC data for CLs is carried out in accordance with the requirements of ISO/IEC 17043 [15]. During of the evaluation of primary data from the participating CLs, the interlaboratory deviation of the measurement results or degree of equivalences (DoE) was calculated based on the ILCs results.

The DoE for j-th CLs participant of ILC is calculated using Equation [4, 34, 38].

$$D_{labj} = x_{labj} - X_{AV}, \tag{1}$$

where x_{labj} is measured value for i-th CL; X_{AV} is AV for ILC.

Expanded uncertainty of the result of each participant $U(x_{labj})$ and expanded uncertainty of AV $U(X_{AV})$ were used to check the consistency of the primary ILC data and to calculate E_n index (E_n number) using equation

$$E_n = D_{labj} / \sqrt{U^2(x_{labj}) + U^2(X_{AV})}. \tag{2}$$

$|E_n| \leq 1.0$ indicates satisfactory performance and $|E_n| > 1.0$ indicates unsatisfactory performance [4, 15].

On **Figures 1–3** show the traditional graphical interpretation of the results of three ILCs at one of the calibration points (ILC 2–1, **Figure 1**, ILC 4–2, **Figure 2** and ILC 6–1, **Figure 3** respectively). The evaluation of primary data of all ILCs is carried out by means of the specially developed software "Interlaboratory comparisons" (Ukraine) which implements the algorithm presented in [34]. To prepare reports on ILCs, RL used specified software that allowed calculating the E_n and z indexes and constructing a graphical display of the results. The figures show the DoE with expanded uncertainty for all participating CLs in ILCs. The green dashed line shows the measurement uncertainty limits of the AV of ILC.

Only two laboratories (lab 4 and lab 6 for ILC 2–1) have an unsatisfactory result for two ILCs using the E_n index. E_n index more characterizes the reliability of measurement results of laboratories participating in the ILC, but is not always sufficient to determine the accuracy of measurement results.

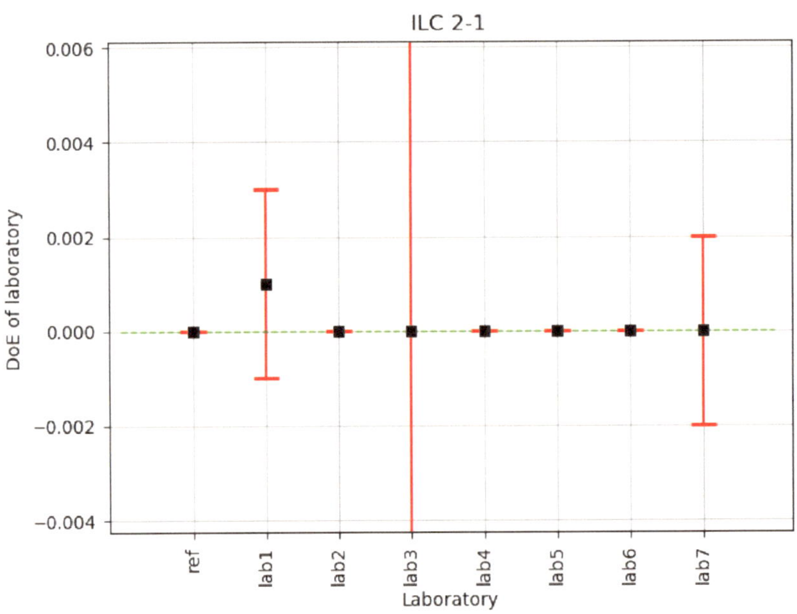

Figure 1.
DoE of CLs for ILC 2–1.

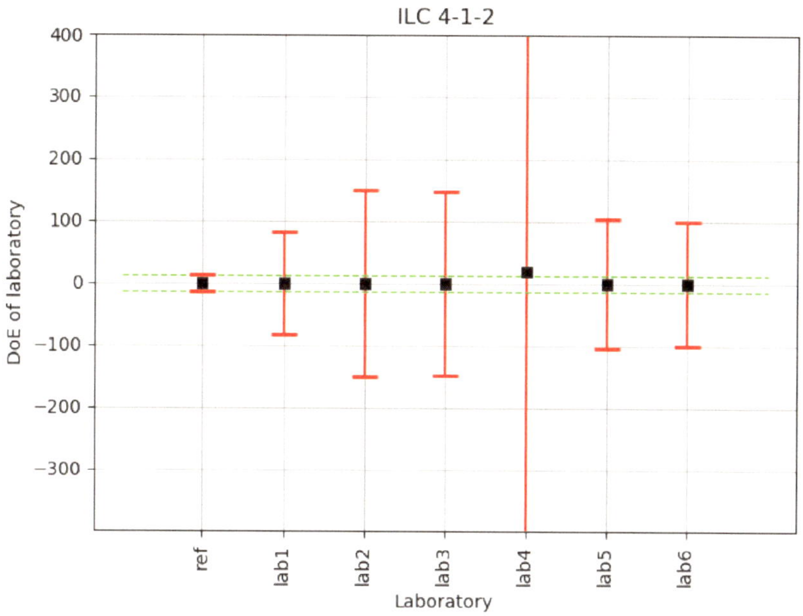

Figure 2.
DoE of CLs for ILC 4–2.

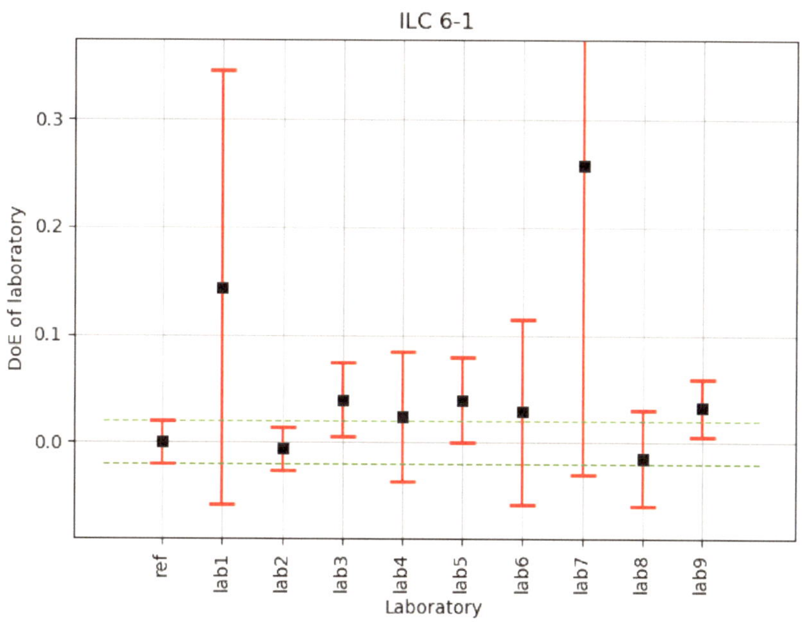

Figure 3.
DoE of CLs for ILC 6–1.

4. The additional data evaluation of interlaboratory comparisons

The consistency evaluation of data using E_n and z indicators is important not only to confirm the technical competence of laboratories participating in the ILC. This will also help to increase the accuracy of calibration by the laboratory participating in the ILC with a corresponding reduction in measurement uncertainty.

The z index compares the measurement results of all laboratories with each other and gives better information about the accuracy of measurements in laboratory. The measurement accuracy is an important characteristic for CL, therefore this index is more suitable for evaluating ILC data for CLs.

z index (z score) is calculated by the equation

$$z = D_{labj}/\sigma, \quad (3)$$

where σ is the standard deviation for qualification assessment (ILC).

$|z| \leq 2.0$ indicates a satisfactory performance characteristic and does not require adjustment or response measures, $2.0 < |z| < 3.0$ indicates a dubious performance characteristic and requires precautionary measures, and $|z| \geq 3.0$ indicates an unsatisfactory performance characteristic and requires adjustment or response measures.

In **Tables 2–8** shows the calculated results of E_n and z indexes at all points of the calibration for all ILCs. E_n and z indexes are zero for RL. Cells with unsatisfactory results are highlighted in grey in the tables. An unsatisfactory result is the excess for E_n index of the value 1, and for the z index of 2 (does not require adjustment or response measures) or 3 (requires adjustment or response measures) [4, 15].

ILC data	Index	Lab 1	Lab 2	Lab 3	Lab 4
ILC1–1	E_n	−0.854	0.444	0.438	−0.312
	z	−0.584	0.114	0.224	−2.462
ILC1–2	E_n	−0.451	0.818	1.522	−0.064
	z	−1.146	0.266	0.605	−2.090
ILC1–3	E_n	−0.645	0.882	0.987	0.022
	z	−1.167	0.483	0.781	1.895
ILC1–4	E_n	−0.147	0.238	0.129	0.452
	z	−0.025	0.019	0.018	2.503
ILC1–5	E_n	—	0.753	0.382	—
	z	—	2.265	1.939	—

Table 2.
Results of ILCs for calibration of precision measuring thermocouple.

ILC data	Index	Lab 1	Lab 2	Lab 3	Lab 4	Lab 5	Lab 6	Lab 7
ILC2–1	E_n	0.249	−0.245	−0.008	−4.631	−0.100	2.352	−0.021
	z	2.677	−0.027	−0.996	−0.208	−0.005	0.188	−0.169
ILC2–2	E_n	−1.955	6.066	−0.018	−0.814	0.256	−0.646	−0.032
	z	−2.404	0.880	−1.460	−0.197	0.056	−0.253	−1.460
ILC2–3	E_n	−0.899	4.179	−0.086	0.171	−0.470	0.975	−0.058
	z	−2.251	1.318	−0.969	0.051	−0.127	0.367	−0.969

Table 3.
Results of ILCs for calibration of measures of electrical resistance (1-th round).

On **Figure 4** shows the graphical interpretation of the results of estimation of E_n (a) and z (b) indexes for ILC 2–1, on **Figure 5** – for ILC 4–2, and in **Figure 6** – for ILC 6–1, respectively.

ILC data	Index	Lab 1	Lab 2	Lab 3	Lab 4
ILC3–1	E_n	0.301	0.133	0.367	0.322
	z	0.117	2.658	0.152	0.961
ILC3–2	E_n	0.194	0.065	0.042	0.051
	z	0.012	2.579	0.004	0.457
ILC3–3	E_n	0.301	0.133	0.367	0.322
	z	0.117	2.658	0.152	0.961

Table 4.
Results of ILCs for calibration of measures of electrical resistance (2-th round).

ILC data	Index	Lab 1	Lab 2	Lab 3	Lab 4	Lab 5	Lab 6
ILC4–1	E_n	0.000	0.004	0.001	−0.013	0.000	0.000
	z	0.002	0.215	0.029	−2.804	0.018	0.018
ILC4–2	E_n	0.000	0.002	0.002	0.034	0.000	0.000
	z	0.001	0.046	0.046	2.875	0.004	0.006
ILC4–3	E_n	0.000	0.002	0.001	0.000	0.000	0.000
	z	0.157	2.683	0.679	−0.192	−0.105	0.157
ILC4–4	E_n	0.000	0.002	0.001	0.000	0.000	0.000
	z	0.207	2.657	1.055	−0.170	−0.075	0.207
ILC4–5	E_n	0.000	0.002	−0.001	0.119	0.000	0.000
	z	0.001	0.013	−0.009	2.859	0.000	0.001
ILC4–6	E_n	0.000	0.002	0.002	0.137	0.000	0.000
	z	0.000	0.012	0.012	2.862	0.000	0.001
ILC4–7	E_n	0.000	0.003	0.003	0.000	0.000	0.000
	z	−0.043	2.260	1.919	−0.043	0.000	−0.043
ILC4–8	E_n	0.000	0.001	0.001	0.000	0.000	0.000
	z	0.176	2.439	2.187	0.176	0.553	0.176
ILC4–9	E_n	0.001	0.002	0.000	0.000	0.000	0.001
	z	0.532	2.787	0.622	−0.009	−0.099	0.532
ILC4–10	E_n	0.000	0.002	0.000	0.000	−0.001	0.000
	z	0.108	2.570	0.569	−0.277	−0.354	0.108
ILC4–11	E_n	0.001	0.002	0.002	0.000	0.000	0.001
	z	0.331	2.462	2.068	0.095	0.253	0.331
ILC4–12	E_n	0.001	0.002	0.001	0.000	0.000	0.001
	z	0.498	2.989	1.329	0.406	0.406	0.498

Table 5.
Results of ILCs for calibration of precision measure of electric power.

ILC data	Index	Lab 1	Lab 2	Lab 3
ILC5–1	E_n	−0.235	−0.014	−0.305
	z	−2.479	−0.077	−1.044
ILC5–2	E_n	−0.050	−0.056	0.036
	z	−1.095	−1.461	1.095
ILC5–3	E_n	0.040	0.000	0.035
	z	1.206	0.000	2.412
ILC5–4	E_n	0.074	0.033	0.068
	z	1.414	1.414	2.828
ILC5–5	E_n	0.737	0.397	0.139
	z	2.399	2.181	0.727
ILC5–6	E_n	−0.087	−0.055	−0.016
	z	−2.557	−1.627	−0.465
ILC5–7	E_n	0.289	0.278	0.122
	z	2.448	2.292	1.011
ILC5–8	E_n	−0.086	0.016	−3.258
	z	−0.063	0.012	−2.326
ILC5–9	E_n	−0.284	−0.061	0.029
	z	−2.315	−0.489	0.233
ILC5–10	E_n	−0.947	−1.024	0.692
	z	−1.335	−1.442	0.975

Table 6.
Results of ILCs for calibration of low frequency signal generator.

ILC data	Index	Lab 1	Lab 2	Lab 3	Lab 4	Lab 5	Lab 6	Lab 7	Lab 8	Lab 9
ILC6–1	E_n	0.713	−0.212	0.992	0.379	0.894	0.328	0.897	−0.290	0.982
	z	1.811	−0.075	0.503	0.302	0.503	0.365	3.245	−0.176	0.415
ILC6–2	E_n	0.998	−0.943	0.733	0.218	0.676	0.161	0.192	−0.379	0.914
	z	3.160	−0.334	0.486	0.729	0.790	0.501	1.686	−0.273	1.413
ILC6–3	E_n	0.711	−0.587	0.459	0.157	0.296	0.065	0.804	−0.273	0.754
	z	2.008	−0.193	0.468	0.908	0.468	0.289	3.012	−0.165	1.582

Table 7.
Results of ILCs for calibration of electronic stopwatch.

ILC data	Index	Lab 1	Lab 2	Lab 3	Lab 4	Lab 5	Lab 6	Lab 7	Lab 8	Lab 9	Lab 10
ILC7–1	E_n	0.050	0.662	−0.072	−0.025	0.643	−0.164	−0.384	−0.039	−0.012	−0.124
	z	1.243	2.666	−0.687	−0.634	0.857	−0.124	−0.588	−0.631	−0.118	−0.433
ILC7–2	E_n	0.067	0.912	−0.076	−0.034	0.775	−0.253	−0.475	−0.068	0.005	−0.052
	z	1.261	2.533	−0.695	−0.629	0.761	−0.182	−0.725	−1.040	0.034	−0.145
ILC7–3	E_n	0.062	1.134	−0.072	−0.053	0.164	−0.160	−0.648	−0.042	−0.014	−0.104
	z	1.045	2.847	−0.746	−0.834	0.164	−0.131	−0.574	−0.427	−0.09	−0.276

Table 8.
Results of ILCs for calibration of high-frequency signal generator.

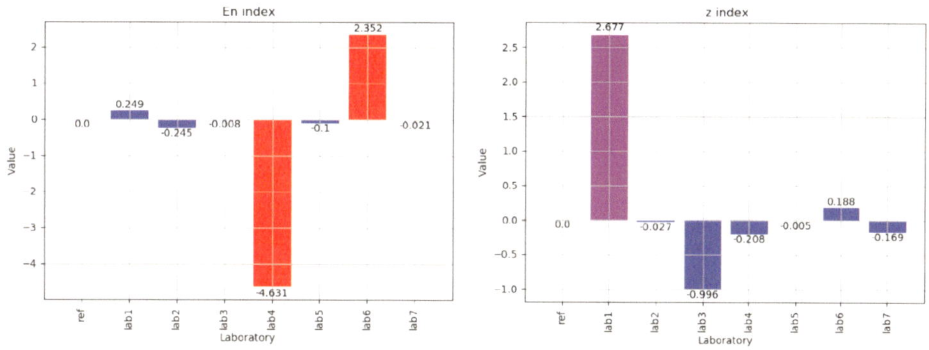

a *b*

Figure 4.
Values of E_n and z indexes for ILC 2–1: a is E_n index, b is z index.

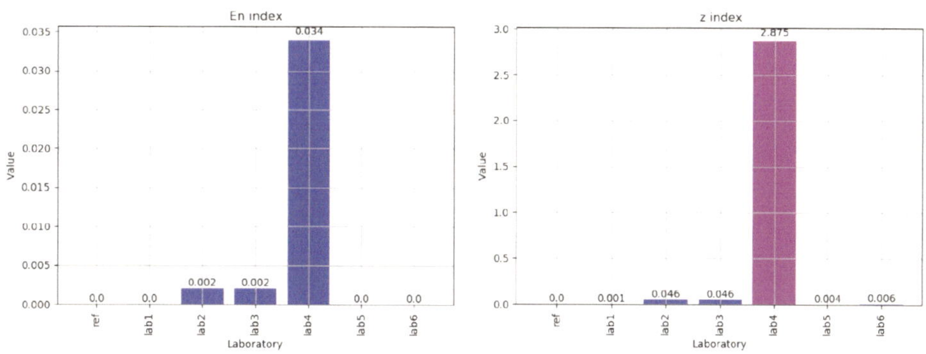

a *b*

Figure 5.
Values of E_n and z indexes for ILC 4–2: a is E_n index, b is z index.

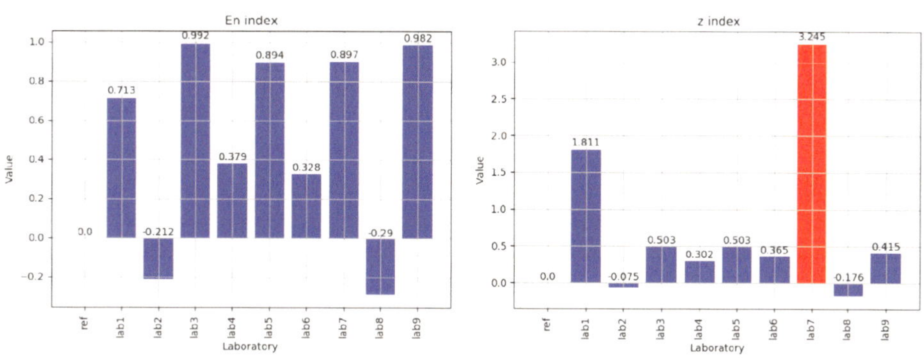

a *b*

Figure 6.
Values of E_n and z indexes for ILC 6–1: a is E_n index, b is z index.

5. The summarized results of interlaboratory comparisons

The summarized results of estimation of E_n and z indexes for all ILCs are shown in **Table 9** and **Figure 7**. The percentage of discrepancies two assessments for ILCs

ILC	Number of participants*	Number of parameters	$E_n > 1$ for lab	$z > 3$ for lab	$z > 2$ for lab	Percentage of discrepancies in evaluation
ILC1	4 labs	20 points	1 point (5%)	0 point (0%)	4 points (20%)	100%
ILC2	7 labs	21 points	5 points (24%)	0 point (0%)	3 points (14%)	95%
ILC3	4 labs	12 points	0 point (0%)	0 point (0%)	3 points (25%)	100%
ILC4	6 labs	72 points	0 point (0%)	0 point (0%)	14 points (19%)	100%
ILC5	3 labs	30 points	2 points (7%)	0 point (0%)	10 points (33%)	97%
ILC6	9 labs	27 points	0 point (0%)	3 points (11%)	1 point (4%)	100%
ILC7	10 labs	30 points	1 point (3%)	0 point (0%)	3 points (10%)	97%
	Total:	**212 points**	**8 points (4%)**	**3 points (1,4%)**	**38 points (18%)**	—

*Without RL.

Table 9.
The summarized results of estimation of En and z indexes for all ILCs.

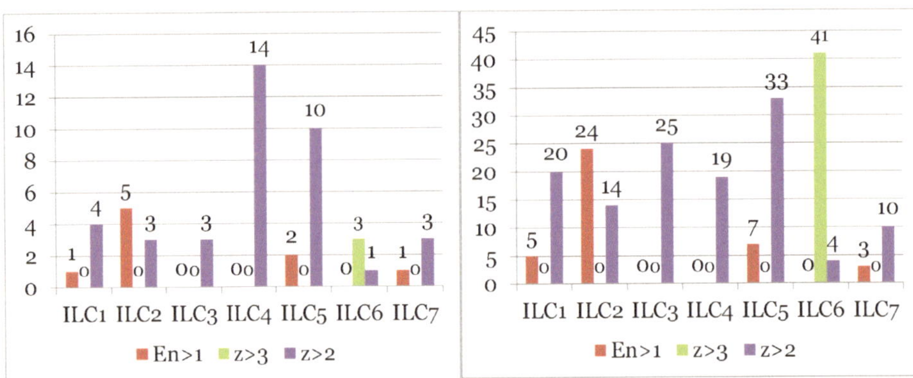

a *b*

Figure 7.
The summarized results of estimation of E_n and z indexes for all ILCs: a is absolute value, b is percentage value (%).

1, 3, 4 and 6 estimates are 100, for ILCs 2, 5 and 7 estimates are from 95 to 97. This suggests that the conclusions that can be drawn about the technical competence of the laboratories participating in these ILCs are completely inconsistent.

Only one result of ILC1 according to E_n index have inconsistency (lab 3). At the same time, 4 results of ILC1 according to z indexes have inconsistencies (lab 2 and lab 4).

5 results of ILC2 according to E_n indexes have inconsistencies (for labs 1, 2, 4 and 6). At the same time, 3 results of ILC2 according to z indexes have inconsistencies (only for lab 1).

ILC3, ILC4, and ILC6 according to E_n index have no inconsistencies. At the same time, 3 results of ILC3 according to z indexes have inconsistencies (only for lab 2), 14 results of ILC4 according to z indexes have inconsistencies (for labs 2, 3 and 4), and 4 results of the ILC6 according to z indexes have inconsistencies (for labs 2, 3, and 4), including 3 from 4 are very large ($z > 3.0$).

Only one ILC7 result according to E_n index have inconsistency (for lab 2). At the same time, 3 results of the ILC7 according to z indexes have inconsistencies (for lab 2 also).

The results of the data consistency analysis show that all ILCs, taking into account both indexes, have measurement points with unsatisfactory results. Analysis of the data taking into account the E_n index shows that only three ILCs (ILC3, ILC4, and ILC6) have satisfactory results. At the same time, analysis of the data taking into account the z index ($z > 2$) shows that all ILCs have measurement points with unsatisfactory results. ILC6 has measurement points with significantly unsatisfactory results, taking into account the z index ($z > 3$).

If we return to the analysis of **Figures 1–3**, it can be seen that lab 4 for ILCs 2–1 and lab 1 and lab 7 for ILC6–1 have very large declared measurement uncertainties with large DoEs. This led to unsatisfactory results, taking into account the z index. The main reason for the unsatisfactory result of lab 3 for ILCs 1–2, taking into account E_n index, is, on the contrary, a very small declared measurement uncertainty.

The general recommendation for lab 3 and lab 4 for ILC1–2, as well as for lab 4 for ILC4–2, and lab 1 and lab 7 for ILC6–1 is to revise the estimate of the measurement uncertainty, taking into account guides [8, 31]. This measurement uncertainty can be influenced by both the calibration results of the laboratory working standards and the level of competence of the laboratory personnel. Taking these recommendations into account can improve the results of that laboratories participation in other rounds of ILCs or new ILCs.

6. The influence of travelling standards instability

The travelling standards instability can affect the results of ILCs for CLs. Some works are devoted to assessing its influence, in particular compensation for its instability. The repeatability of a good measuring instrument is below 10% of its maximum error as shown in [39]. The travelling standard with 0,2% shows variations of random errors below $\tilde{x} \pm 0.02\%$ where $\tilde{x}$ is the average of the readings during calibration. This a small Type A uncertainty in relation to other components is show.

Typically, RL already takes into account the travelling standards instability in the ILC assigned value X_{AV} and its expanded uncertainty [4].

$$U(X_{AV}) = 2 \cdot \sqrt{u^2(x_{ref}) + u^2(x_{inst})}, \tag{4}$$

where $u(x_{ref})$ is the standard measurement uncertainty obtained by calibrating of travelling standard with a RL; $u(x_{inst})$ is the standard measurement uncertainty from the travelling standard instability of during ILC period

$$u(x_{inst}) = \Delta X_{\max}/\sqrt{3}, \tag{5}$$

$\Delta X_{\max}$ is the maximum change in nominal value of travelling standard during ILC period.

The absence of a significant effect of the travelling standards instability on the evaluation of the CL result in the ILC can be at its maximum instability, which is determined by the expression [39].

$$x_{inst} \leq \sqrt{U^2(x_{labj}) + U^2(X_{AV})}. \tag{6}$$

The value of the travelling standards instability can be obtained for several cases: measurements of the RL of the travelling standard in the process of carry out of ILC;

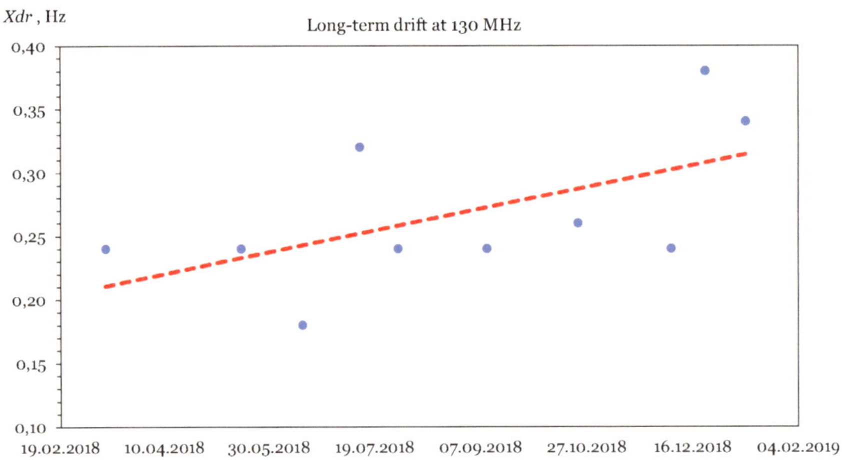

a

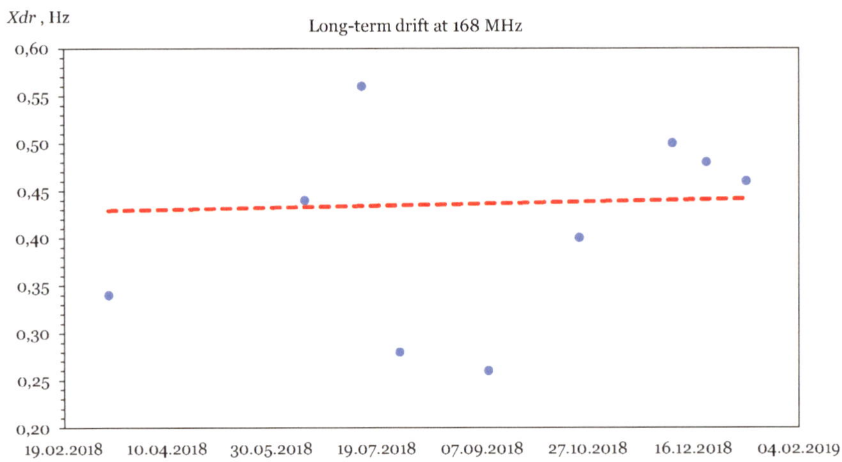

b

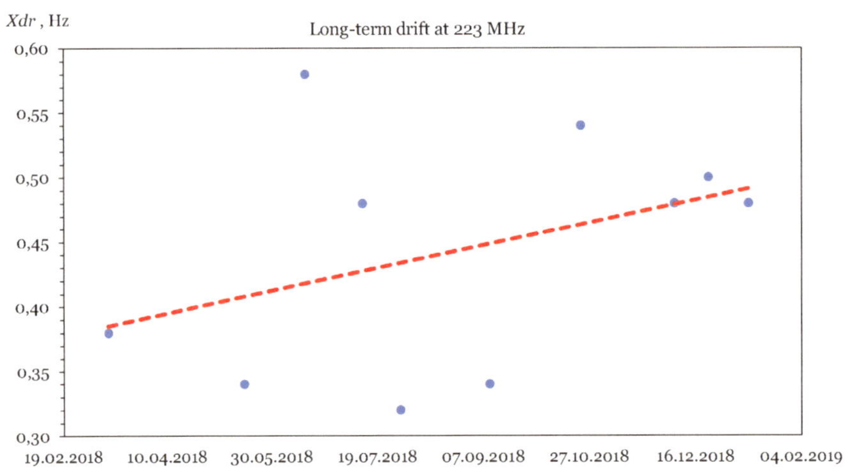

c

Figure 8.
The drift of travelling standard for ILC7: a is frequency 130 MHz, b is frequency 168 MHz, c is frequency 223 MHz.

from the technical specification for the travelling standard, measurements of the RL of the travelling standard for a long time, and etc. Results of calculating the E_n index for various options for accounting for travelling standards instability are shows in [39].

In any case, from expression (4) it follows that with an increase in the value of the measurement uncertainty associated with instability, the value of the E_n index only decreases. In this case, it can be stated that the use of a more unstable travelling standard can improve the consistency of the ILC data, which is not acceptable for CL. To carry out ILCs for CLs, it is more preferable to use working standards as calibration objects. Typically, a working standard has less instability than a measuring instrument. The use of a measuring instrument as a travelling standard can lead to somewhat distorted results of such ILCs.

An analysis RL of the travelling standard instability for all calibration points of the ILC7 is given in [33]. The drift of travelling standard for ILC7 at all frequencies is presented on **Figure 8**. The uncertainty of travelling standards instability for ILC7 is presented in **Table 10**. The contribution of the uncertainty from the long-term drift of the travelling standard to the standard uncertainty of AV for the entire duration of ILC7 is from 5.3 to 8.3% for all calibration points. Such a drift of the measuring instrument used as a calibration object is acceptable for the ILC. It does not distort the ILC results for the participating CLs.

The list of travelling standard for all ILCs and values of E_n and z indexes are shown in **Table 11.**

The use of a measuring instrument as a calibration object leads to a slight increase in the values of the z index and practically does not affect the E_n index, as can be seen from **Table 11**.

ILC7 point	Frequency (MHz)	$u(X_{AV})$ (Hz)	$u(x_{inst})$ (Hz)	Drift contribution to uncertainty AV (%)
ILC7–1	130	0.12	0.01	8.3
ILC7–2	168	0.17	0.01	5.9
ILC7–3	223	0.19	0.01	5.3

Table 10.
The uncertainty of travelling standards instability for ILC7.

ILC	Calibration object	Working standard	Measuring instrument	E_n index	z index
				Unsatisfactory (%)	
ILC1	Precision measuring thermocouple	Yes	No	5	20
ILC2	Measures of electrical resistance (1th round)	Yes	No	24	14
ILC3	Measures of electrical resistance (2th round)	Yes	No	0	25
ILC4	Precision measure of electric power	Yes	No	0	19
ILC5	Low frequency signal generator	No	Yes	7	33
ILC6	Electronic stopwatch	No	Yes	0	45
ILC7	High-frequency signal generator	No	Yes	3	13

Table 11.
The list of travelling standard for ILCs and values of E_n and z indexes.

7. The improvement of the evaluation of interlaboratory comparison results

Statistical methods for use in proficiency testing by ILCs are presented in [26, 27]. The aim of creating alternative statistics in order to improve the analysis and evaluation of ILC measurement results is research work [40]. The improvement of statistical indicators is proposed by addressing two specific issues: robustness and reliability. The proposed methodology is not traditional for ILC, but it can be used as an additional methodology for checking the results of ILC.

The following conditions are provided for data evaluation of international comparison of national standards: the travelling measurement standard is stable, the measurement results presented by laboratories are reciprocally independent, and the Gaussian distribution is assigned to a measurand in each laboratory [41–44]. The same conditions can be extended for data evaluation of ILCs for CLs. Frequently the measurement procedures for supplementary comparisons of national standards [44] are the calibration procedures of these laboratories. Such calibration procedures can also be extended to ILCs for CLs. In such a case, the calibration capabilities of the laboratory can be confirmed.

The application of z index for evaluation of CL results recommended instead of E_n index since this number is not applicable due to the difficulty in determining the AV [16]. Of course, for accredited test laboratories, it is preferable to use the services of a CL with the best calibration capabilities. Better calibration capabilities of laboratories are characterized by lower calibration uncertainties of working standards and measuring instruments. CLs with a satisfactory value of the E_n index in the ILC, but having large calibration uncertainties become uncompetitive. If we return to the analysis of **Figures 1–3**, it can be seen that lab 1, 3 and 7 for ILCs 1–2, lab 4 for ILC4–2, and lab 1 and lab 7 for ILC6–1 have very large declared measurement uncertainties.

The declared measurement uncertainties of CL for ILC are judged as confirmed if the following equation is satisfied [4, 43].

$$|D_{labj}| < 2u(D_{labj}). \tag{7}$$

In case the declared uncertainties CL don't confirmed during the ILC and for their confirmation it is necessary to participate in other similar ILCs.

Often, a national metrological institute or an accredited CL, which is an RL in ILC, performs high-precision calibration of working standards and measuring instruments for CLs participating in this ILC. In this case, a correlation of the obtained CL results is formed, which must be taken into account when evaluating the data of such an ILC. Covariance's are estimated by careful analysis of the uncertainty budget of CLs by the RL

$$\mathrm{cov}(x_{labj}, X_{AV}) = u_0^2, \tag{8}$$

where u_0^2 is common input to the uncertainty budgets of both results [43].

In this case, the value of the E_n index is calculated by the formula:

$$E_n = D_{labj} / \sqrt{U^2(x_{labj}) + U^2(X_{AV}) - 2\mathrm{cov}(x_{labj}, X_{AV})}. \tag{9}$$

If the value of the E_n index meets the specified requirement (≤ 1.0), then the minimum standard measurement uncertainty, that can be claimed as calibration capability of CL participating in ILC, is:

$$u_{CC}(x_{labj}) = u_{ILC}(x_{labj}). \quad (10)$$

If the value of the E_n index not meets the specified requirement (> 1.0), then the minimum standard measurement uncertainty, that can be claimed as calibration capability of CL participating in ILC, is:

$$u_{CC}(x_{labj}) = \sqrt{\frac{D_{labj}^2}{4} + u^2(X_{AV})}, \quad (11)$$

where $u(X_{AV})$ is the standard measurement uncertainty of AV.

Correspondingly, the extended uncertainty is $U_{0.95}(x_{labj}) = 2u_{ILC}(x_{labj})$.

The same requirements can be extended for compliance (≤ 2.0) or inconsistency (> 2.0) of the value of the z index with the established requirements. In this case, the minimum standard measurement uncertainty, that can be claimed as calibration capability of CL participating in ILC, will be determined by formulas (10) and (11), respectively.

If the standard uncertainty $u(X_{AV})$ of the AV is too large in comparison with the standard deviation σ for ILC, then there is a risk that some laboratories will receive action and warning signals because of inaccuracy in the determination of the AV, not because of any cause within the laboratories. If

$$u(X_{AV}) \leq 0.3\sigma \quad (12)$$

then the uncertainty of the AV is negligible and need not be included in the interpretation of the ILC results. Further, all CLs participating in ILC shall carry out the same number of replicate measurements. This approach assumes that CLs have generally similar repeatability [26].

To evaluate the ILC data, can use z' index also [26] which calculated by the equation

$$z' = D_{labj} / \sqrt{\sigma^2 + u^2(X_{AV})} \quad (13)$$

This equation may be used when the AV is not calculated using the results reported by CLs participating in ILC. z' index shall be interpreted in the same way as z index and using the same critical values of 2.0 and 3.0.

Comparison of the equations for z and z' indexes shows that z' index for ILC will all be smaller than the corresponding z index by a constant factor of $\sigma / \sqrt{\sigma^2 + u^2(X_{AV})}$.

When the inequality established by expression (12) is satisfied, then this factor will fall in the range: $0.96 \leq \sigma / \sqrt{\sigma^2 + u^2(X_{AV})} \leq 1.00$. In this case, z' index will be nearly identical to z index, and it may be concluded that the uncertainty of the AV is negligible. When the inequality established by expression (12) is not satisfied, the difference in magnitude of the z' and z indexes may be such that some z index exceeds the critical values of 2.0 or 3.0.

8. Conclusions

To perform an ILC for CLs, RL must provide a stable working standard or measuring instrument as a calibration object and monitor its drift throughout the ILC. The use of a measuring instrument as a calibration object leads to a slight increase in the values of the z index and practically does not affect the E_n index. The application of z index for evaluation of CL results recommended instead of E_n index.

The analysis of the results of the ILC for CLs for consistency should include not only the analysis of the values of the E_n index, but also the z index. If we restrict ourselves to only the E_n index, then it is possible to get unreliable results of the ILC and not identify problems in the CL-participants of the ILC. In 3 from 7 ILCs examined, the E_n index showed completely satisfactory results, while the z index in all of these 3 ILCs revealed problematic results from the participating laboratories.

The stable travelling standard, the independent measurement results of laboratories with Gaussian distribution are main conditions for data evaluation of ILC for CLs. To participate in the ILC when declaring its measurement uncertainty, CLs must conduct a thorough analysis of the components of this uncertainty. It is necessary to take into account the correlation of the laboratory data of the participants of the ILC when evaluating its results. Covariance is estimated by carefully analyzing the CL uncertainty budget using RL.

The minimum standard measurement uncertainty that can be claimed as the calibration capability of a CL participating in an ILC can be determined in different ways depending on the value of the obtained E_n index or z index. If the standard uncertainty of AV is too large compared to the standard deviation for the ILC, there is a risk of unreliable results for some CLs.

Author details

Oleh Velychko[1]* and Tetyana Gordiyenko[2]

1 State Enterprise "Ukrmetrteststandard", Kyiv, Ukraine

2 State University of Telecommunications, Kyiv, Ukraine

*Address all correspondence to: velychko@ukrcsm.kiev.ua

References

[1] ILAC B7:10/2015. The ILAC Mutual Recognition Arrangement, ILAC, 2015; 8 p.

[2] ILAC P4:05/2019. ILAC Mutual Recognition Arrangement: Policy and Management, ILAC, 2019, 11 p.

[3] ILAC P5:05/2019 ILAC Mutual Recognition Arrangement: Scope and Obligations, ILAC; 2019, 9 p.

[4] Velychko O., Gordiyenko T. Metrological Traceability at Different Measurement Levels. Standards, Methods and Solutions of Metrology, Published by IntechOpen, London, United Kingdom, 2019; Chapter 1, pp. 1–21. DOI: 10.5772/intechopen.84853.

[5] ILAC P9:06/2014 ILAC Policy for Participation in Proficiency Testing Activities, ILAC; 2014, 8 p.

[6] ILAC P10:07/2020. ILAC Policy on Traceability of Measurement Results, ILAC; 2020, 11 p.

[7] ILAC P14:09/2020 ILAC Policy for Measurement Uncertainty in Calibration, ILAC; 2020, 14 p.

[8] JCGM 100:2008. Evaluation of measurement data – Guide to the expression of uncertainty in measurement (GUM), JCGM; 2008. 134 p.

[9] JCGM 100:2008. Evaluation of measurement data — Supplement 1 to the “Guide to the expression of uncertainty in measurement” – Propagation of distributions using a Monte Carlo method, JCGM; 2008. 90 p.

[10] JCGM 104:2009. Evaluation of measurement data – An introduction to the “Guide to the expression of uncertainty in measurement” and related documents, JCGM; 2009. 28 p.

[11] JCGM GUM-6:2020. Guide to the expression of uncertainty in measurement – Part 6: Developing and using measurement models, JCGM; 2020. 103 p.

[12] ILAC–BIPM Partnership [Internet]. Available from: https://ilac.org/about-ilac/partnerships/international-partners/bipm/ [Accessed: 2021-06-25].

[13] ILAC G17:01/2021. ILAC Guidelines for Measurement Uncertainty in Testing, ILAC; 2021, 12 p.

[14] ISO/IEC 17025:2017. General requirements for the competence of testing and calibration laboratories. Switzerland: ISO/IEC; 2017. 30 p.

[15] ISO/IEC 17043:2010. Conformity assessment. General requirements for proficiency testing. Switzerland: ISO/IEC; 2010. 39 p.

[16] Beckert S. F., Fischer G. E. Interlaboratory comparison of roughness measurement: Application of Algorithm A of ISO 13528:2015 in determining the designated value and the standard deviation. XXII World Congress of the International Measurement Confederation (IMEKO 2018). Journal of Physics: Conf. Series. IOP Publication. 2018;1065;082007:4. DOI: 10.1088/1742-6596/1065/8/082007.

[17] Iacobescu F., Poenaru M. M., Anghel M.-A. Reactive Power Quality Assessment through Interlaboratories Comparison. 22th IMEKO TC 4 Symposium “Supporting World development through electrical and electronic measurements”, 2017; Iasi, Romania, 13–19.

[18] Poenaru M. M., Iacobescu F., Anghel M.-A. Length calibration Quality assessment through Interlaboratories Comparison. 22th IMEKO TC 4

Symposium “Supporting World development through electrical and electronic measurements”, 2017; Iasi, Romania, 20–26.

[19] Poenaru M. M., Iacobescu F., Anghel M.-A. Pressure Calibration Quality Assessment through Interlaboratories Comparison. 22th IMEKO TC 4 Symposium “Supporting World development through electrical and electronic measurements”, 2017; Iasi, Romania, 27–32.

[20] Poenaru M. M., Iacobescu F., Anghel A.-C., Salceanu A., Anghel M.-A. Active power quality assessment through interlaboratories comparison, 21st IMEKO TC4 International Symposium and 19th International Workshop on ADC Modelling and Testing Understanding the World through Electrical and Electronic Measurement, Budapest, Hungary, 2016, 224–228.

[21] Furuichi N., Terao Y., Ogawa S., Cordova L., Shimada T. Inter-laboratory comparison of small water flow calibration facilities with extremely low uncertainty. Measurement, 2016;91: 548–556.

[22] Bermanec L. G., Zvizdic D. Interlaboratory comparison in the pressure range from 0 to 2 MPa for accredited calibration laboratories. Int. J. Metrol. Qual. Eng., 2015;6, 307. DOI: 10.1051/ijmqe/2015021.

[23] Claudio J., Costa M. Brazilian energy interlaboratory program applicative. In: Proceedings of the XX IMEKO World Congress “Metrology for Green Growth”; 2012; Busan, Republic of Korea. IMEKO; 2012. 6 p.

[24] Briggs P. Proficiency testing for calibration laboratories. In: Proceedings of the XX IMEKO World Congress “Metrology for Green Growth”; 2012; Busan, Republic of Korea. IMEKO; 2012. 5 p.

[25] Velychko O., Gordiyenko T. The estimation of the measurement results with using statistical methods. Journal of Physics: Conf. Series. 2015; 588; 012017:6. DOI: 10.1088/1742-6596/588/1/012017.

[26] ISO 13528:2015. Statistical methods for use in proficiency testing by interlaboratory comparisons. Switzerland: ISO; 2015. 89 p.

[27] Velychko O., Gordiyenko T. The implementation of general guides and standards on regional level in the field of metrology. Journal of Physics: Conf. Series. 2010; 238;012044:6. DOI: 10.1088/1742-6596/238/1/012044.

[28] Chunovkina A., Zviagin N., Burmistrova N. Interlaboratory comparisons. Practical approach for data evaluation. In: Proceedings of the XX IMEKO World Congress “Metrology for Green Growth”; 2012; Busan, Republic of Korea. IMEKO; 2012. 5 p.

[29] Jackson G. S., Muzikar P., Goehring B. A Bayesian approach to an interlaboratory comparison. Chemometrics and Intelligent Laboratory Systems, (2015) Vol. 141, 94–99

[30] Acko B., Brezovnik S., Sluban B. Verification of Software Applications for Evaluating Interlaboratory Comparison Results. Procedia Engineering. 2014;69, 263–272.

[31] EA-04/02 M. Evaluation of the Uncertainty of Measurement in Calibration. EA; 2013. 75 p.

[32] Velychko O., Gordiyenko T., Boriak K. Linking the Results of Inter-laboratory Comparisons for DC Electrical Resistance Measures. Ukrainian Metrological Journal. 2020;1: 4–11. DOI: 10.24027/2306-7039.1.2020.204155.

[33] Velychko O., Shevkun S., Mescheriak O., Gordiyenko T.,

Kursini S. Interlaboratory comparisons of the calibration results of signal generator. Eastern-European Journal of Enterprise Technologies. 2019;3/9(99): 14–20. DOI: 10.15587/1729-4061.2019.166504.

[34] Velychko O., Shevkun S., Gordiyenko T., Mescheriak O. Interlaboratory comparisons of the calibration results of time meters. Eastern-European Journal of Enterprise Technologies. 2018;1/9(91): 4–11. DOI: 10.15587/1729-4061.2018.121089.

[35] Velychko O., Isaiev V. Interlaboratory comparison in context of inappropriate results of voltage thermal converter calibration. Journal of Electrical Engineering and Information Technologies. 2018;3;1–2:5–12.

[36] The BIPM key comparison database (KCDB) [Internet]. Available from: http://kcdb.bipm.org/ [Accessed: 2021-06-25].

[37] Velychko O., Isaiev V. A comparative analysis of AC/DC transfer standards for comparison of national standards. Eastern-European Journal of Enterprise Technologies. Applied physics. 2018;6/5(96):14–24. DOI: 10.15587/1729-4061.2018.150459.

[38] Velychko O., Gordiyenko T. Features of the processing of results and estimation of measurement uncertainty of inter-laboratory comparison for calibration laboratories. Information Processing Systems. 2018;4(155):77–83. DOI: 10.30748/soi.2018.155.10.

[39] Sousa J. J. L., Leitão L. T. S., Costa M. M., Faria M. C. Considerations on the influence of travelling standards instability in an interlaboratory comparison program. In: Proceedings of the XX IMEKO World Congress "Metrology for Green Growth"; 2012; Busan, Republic of Korea. IMEKO; 2012. 4 p.

[40] Berni R., Carobbi C. Alternative statistical analysis of interlaboratory comparison measurement results. In: Proceedings of the XXI IMEKO World Congress "Measurement in Research and Industry"; 2015; Prague, Czech Republic. IMEKO; 2015. 5 p.

[41] Cox M. G. The evaluation of key comparison data: determining the largest consistent subset. Metrologia. 2007;44:187–200.

[42] Cox M. G. The evaluation of key comparison data. Metrologia. 2002;39: 589–595. DOI: 10.1088/0026-1394/39/6/10.

[43] COOMET R/GM/14:2016. Guidelines for data evaluation of COOMET key comparison, COOMET; 2016, 6 p.

[44] COOMET R/GM/19:2016. Guideline on COOMET supplementary comparison evaluation, COOMET; 2016, 5 p.

Chapter 3

New Trends of Optical Measurements

Oleg Angelsky, Peter Maksymyak, Claudia Zenkova, Olexander Ushenko and Jun Zheng

Abstract

Some of the achievements of modern optical metrology are offered for the reader at this chapter. Optical testing methods have always attracted by their important advantages: distance, non-destructive impact on the test object and, of course, high accuracy. So, using of polarization optics approach, the authors were able to implement the methods for controlling the surface roughness for the moving surface with the measurement accuracy of 10 angstroms. It has become possible to make a breakthrough in the basic methods of measurements from the nano to the femto or pico units of the measured quantity value over the past decades. Control of nano (micro) particle motion by an optical field and their use for testing complex optical fields; ultra-precise determination of the optical parameters of both solid and liquid and gas-like substances by optical methods; by interference methods and many other, are proposed for consideration here. Some biomedical applications are also offered for reader's familiarization. Particularly, the results of 3D Stokes-polarimetric mapping of microscopic biological images with digital holographic reproduction of layer-by-layer ellipticity polarization maps in differential diagnosis of benign and malignant tumors with different degrees of differentiation are presented. The authors have shown that using of polarization-holographic measurements in biomedical applications makes it possible to obtain a reliable diagnostic of pathological states both of biological fluids and solid-state objects.

Keywords: micro- (nano)- object, optical force, spin (orbital) momentum, biological image, interference method, measurement accuracy, measured value, error

1. Introduction

Metrology is a field of scientific and technical research that has absorbed optical engineering and precision measurements, directions that intensively develop.

This is proclaimed by the world achievements in the creation of femtosecond lasers (John L. Hall and Theodor W. Hänsch "for their contributions to the development of laser-based precision spectroscopy, including the optical frequency comb technique", Nobel Prize in Physics 2005), the making and successful operation of an optical fluorescence microscope with a record resolution of 10 nanometers (Eric Betzig, Stefan W. Hell and William E. Moerner "for the development of super-resolved fluorescence microscopy", Nobel Prize in Chemistry 2014) [1], and the intensive research in the development and use of optical tweezers, manipulators, molecular motors, etc. (Arthur Ashkin "for the optical tweezers and their application to biological systems", Nobel Prize in Physics 2018).

Using the opportunity of optical diagnostics, observation, and measurement, which are opening up thanks to the development of modern methods for the formation of various features of structured light [2], it has become possible to reach the verge of a picometer level resolution of measurements in a practically revolutionary way.

Already today, ways of implementing the measurement of optical field parameters in three-dimensional space [3–5] with nanometer resolution have been outlined. The appropriateness and importance of such measurements are strengthened in the case of study of polarized light transformations and the possibility of realization an ultra-high resolution in the optical range. In this sense, it is necessary to present some biomedical application and to notice the development and experimental testing of a new 3D Stokes-polarimetry method based on mapping of object fields of biological optically anisotropic layers [6, 7]. Here digital holographic layer – by – layer reconstruction of polarization elliptisity distributions is used, e.g. for express diagnosis and differentiation of samples of polycrystalline blood films of patients with prostate tumors [8–10].

The today's state of physics and life sciences, in general, require expanding the abilities of optical systems by moving from the systems of the so-called millimeter-micrometer range to instruments and devices operating in the nano-, pico-, femto-meter ranges.

The implementation of such tasks is impossible without finding and discovering new technical and technological solutions, the creation of hybrid measurement systems that can simultaneously work in different spectral ranges, using different parameters of the optical radiation field for diagnostics, and evidently their interconnection.

A review of the state of research at this level is based on a platform that has been formed in a number of directions by the authors of this chapter.

So, the goal of this presentation is to analyze new trends of precision measurements in the picometer range, based largely on the results of own research by the authors.

2. Micro and nanostructures and metrology applications

2.1 Using of micro- and nanoparticles for metrology

A new unique tool that has appeared relatively recently and enables optical measurement to be carried out, is a set of methods for capturing, moving, manipulating, controlling of the micro- and nanoparticles motion. To this end, optical traps are created in which optical flows can be controlled using a set of field parameters, namely, amplitude, phase, or polarization.

Thus, various types of traps are formed for spatial capture and movement of nanoscale objects, creating active optical forces of the order of femtonewton [11–13].

Comprehensively considering the methods of creating micro- and nanomanipulators, tweezers and motors, as well as using mey elements, according to our work [11–14], based on classical optical principles, but supplemented with new, and possibly fundamentally new technical and technological solutions for auxiliary devices, new technical problems for optical measurements of micro-, nano - pico femtosecond ranges are solved. Even in the fact that by the speed of microparticles rotation one can determine the absorption coefficient of a matter, the degree of anisotropy of the material, etc. Basically, it can be claimed that these solutions are built on differential, comparative principles and approaches.

2.2 Biaxial crystals in the tasks of creating multifunctional traps for micro- and nano-objects

Based on modern measurement approaches to achieve a nano (femto) units of optical measurements, the possibility for formation of various amplitude-phase, polarization field structures that can be used for manipulating micro- and nano-objects is shown in [12–14]. An interesting solution to this problem is that due to the use of birefringent crystals, it is possible to simultaneously work with various parameters of the optical field, creating a wide arsenal of traps based on one crystal.

Created traps, trapping nanoparticle by an optical field provide determination of nanoparticle parameters with an error of several percent [15].

So the use of birefringent crystals sets the formation of optical fields with deep structure of the inner energy flows. Spread of a slightly divergent (conical) light beam with prescribed linear polarization generates a complicated optical field with spatially inhomogeneous distributions of intensity, phase and polarization that is accompanied by an intricate pattern of the transverse energy flows (**Figure 1**). Here the reader can estimate field distributions at the biaxial crystal output in the vicinity of the optical axis O. Both axes lie in the plane (XZ) so that $x = 0$ corresponds to the middle direction between the axes. The input beam polarization makes an angle 45° with axis X. **Figure 1a** presents both intensity distribution after the output X-oriented polarizer (background) and the transverse orbital flow of the X-polarized component (black arrows). The spin density of the total output field (background), the spin flow map (cyan arrows) and the polarization distribution (gray ellipses) are described in **Figure 1b**.

Such fields offer a variety of possibilities for microparticles' trapping and control, for example:

a. the intensity minima (maxima) form natural traps for absorbing (dielectric) particles due to the gradient force [16, 17];

b. phase singularities associated with the amplitude zeros (**Figure 1**) are coupled with the vortex-like orbital flows able to induce rotation of the trapped particle;

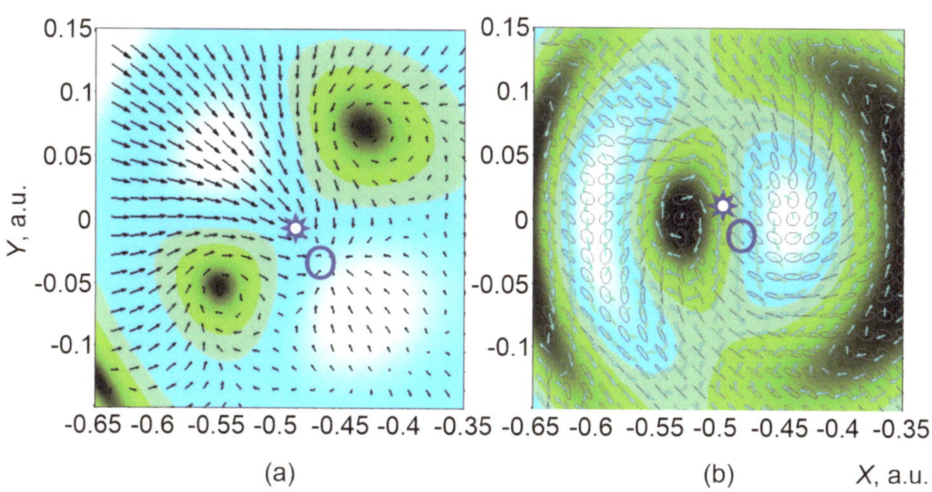

Figure 1.
The scheme of intensity (a), polarization (b) and energy flow (a,b) distribution formed by a birefringent crystal.

c. both the orbital and spin flows can induce definite motion of particles in the transverse plane (**Figure 1a** and **b**);

d. the density of spin angular momentum (**Figure 1b**) may induce controllable orbital motion of particles depending on their position within the beam cross section;

e. the output pattern of optical field distribution can be easily modified via the controllable input and output polarization, providing suitable means for delicate spatial positioning of the trapped particles.

Arrangement of an optical tweezer employing the above principles is shown in **Figure 2**. This optical arrangement based on a birefringent crystal consists from a laser (1) (650 nm, 200 mW), quarter-wave plate – 2, two polarizers – 3, 11, beam expander with spatial filtering – 4, mirror – 5, microobjectives – 6, 8, 13, 17, biaxial crystal with 3D rotating drive – 7, calcite plate – 9, calcite wedges with 2D shifting of one wedge – 10, beam-splitting dichroic cube – 12, sample – 14, white-light source – 15, condenser lenses –16, CCD-camera – 18, Personal Computer (PC) - 19.

On the other hand, as already noted, the use of nano and micro particles provides a diagnostic tool for optical fields. The optical forces arising in the optical field and acting on these particles are at the nano-, pico- femto- Newton, differing in accordance with the properties, shape and size of the particles. Accordingly, it becomes possible to spatially separate internal optical flows, both with respect to spin and orbitally one.

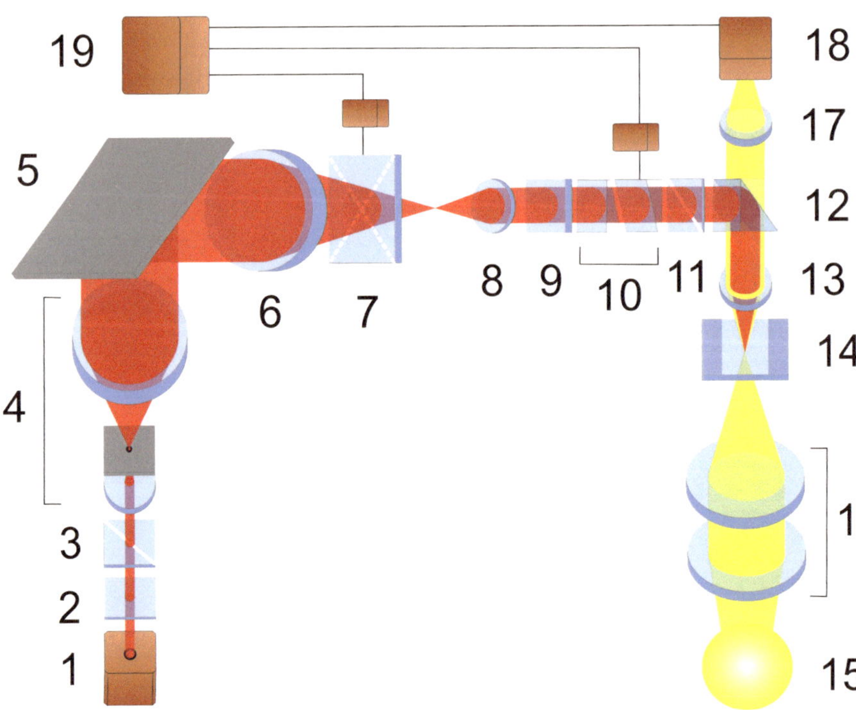

Figure 2.
Optical arrangement of tweezers based on a birefringent crystal.

2.3 Micro and nanoparticles as field probes

Our next achievements [11–14] demonstrate the results obtained by separating the contribution of the orbital and spin angular momentum to the total picture of trapped particles motion in the optical field.

To identify the internal spin energy flows [12], it was necessary to analyze the mechanical action of the spin momentum by testing, selecting size and property of the particles. The spin momentum manifests itself "in its pure form" with all the specific properties in the situation of a symmetric superposition of circularly polarized plane waves and thus the formation of a circularly polarized field having inhomogeneous energy (**Figure 3**) [11, 13]. The mechanical action of the incident field on test particles, when including the scattering components, was carried out through the calculation of the Cartesian components of the force (F_x, F_y, F_z) acting on particles that are placed in the optical field. The longitudinal component of the force (F_z) represents the traditional action of light pressure, which direct the particles forward; the transverse y-component (F_y) corresponds to the gradient force (F_y^{grad}) of an inhomogeneous optical field and traps particles or repels them from areas of high concentration of electromagnetic energy. The most interesting result is the analysis of the component $F_{sp}^{\pm}$ of the optical power along the transverse direction F_x - the only component of the force that is associated with the spin flow. This conclusion is confirmed by the fact that, in accordance with the behavior of the spin flow, the value of the force F_x changes its sign with a change of the helicity σ of the incident beam. In the case of linearly polarized light, this component of the force completely disappears.

Despite the mechanical equivalence of spin and orbital energy flows, that is, their ability to cause translational or orbital motion of the particles under study, the quantitative features of spin-induced and orbital-induced motion, their dependence on the size and properties of the particles are significantly different. That is, the effect of spin and orbital flows can be experimentally distinguished by using test particles with specially selected size and properties. The following figure (**Figure 4**) [13] shows the dependence of the components' value of the optical forces on the particle size for different types of particles. The calculations were performed for two types of spherical particles suspended in water (ε = 1.77, μ = 1, n = 1.33): metal (gold hydrosol, relative refractive index m = 0.32 + 2.65i [18]), dielectric (latex in water, t = 1.12); wave number k = 1.33 · 10^5 cm^{-1} (He-Ne laser).

To eliminate the influence of the incident beam intensity and reduce the number of data presented, the obtained value of the force was normalized by dividing the calculated values by the total momentum of the incident flow over its cross section (P_0). A comparative picture of the mechanical action of the optical forces associated with the spin and orbital internal energy is presented in **Figure 4**. To compare the

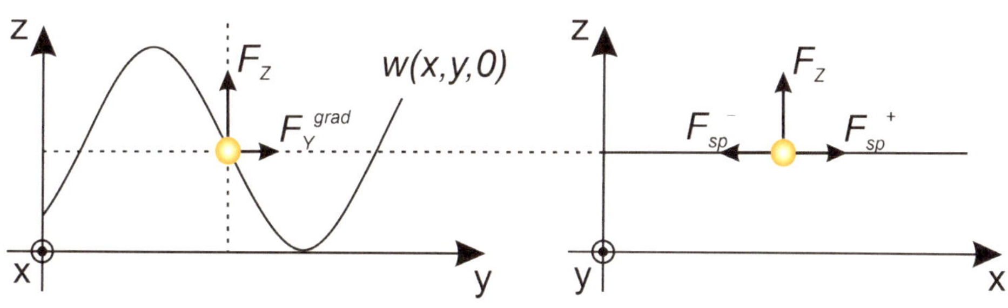

Figure 3.
The components of optical force distribution.

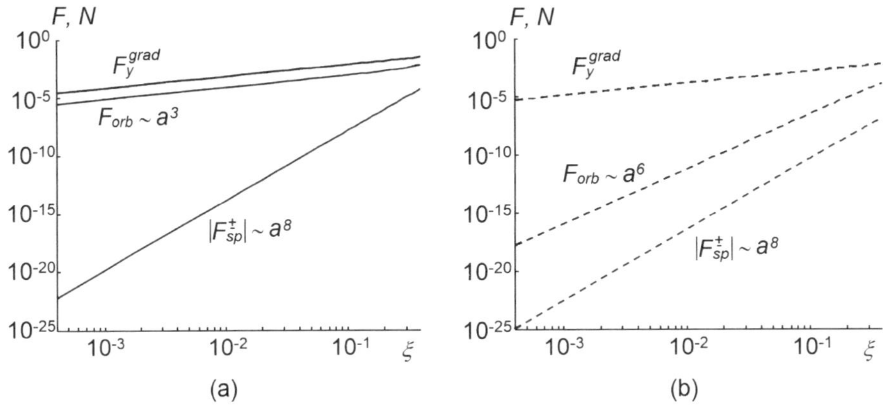

Figure 4.
A comparative picture of the mechanical action of the optical forces associated with the spin and orbital internal energy for different types of particle: a) metal and b) dielectric.

optical force, the gradient force F_y^{grad} is shown. The measured gradient force range is 10^{-4} to 10^{-6} Newton, depends on the particle size and its type and determines the measurement accuracy as 10^{-7} Newton.

The curves are made for particles of small size. Solid lines describe metal particles, dotted lines describe dielectric particles. Here ξ = ka is a dimensionless particle size parameter (*a* – is the particle radius).

As can be seen from the results of estimating the values of spin and orbital flows shown in the figure, the quantitative features of spin-induced and orbital-induced motions and, accordingly, the generated optical forces, their relationship with the size and properties of the particles are significantly different. Following the main aim of this paper, conducting optical power measurement according to the evaluation results (**Figure 4**), we can note the measured quantity of optical force up to 10^{-15}–10^{-25} degree for the spin-induced component of the optical force in accordance with the particle size. We managed to carry out experimental confirmation of the existence of a force of this units of Newton in a specially organized experiment [11–13]. The use of dielectric test particles of the Rayleigh light scattering mechanism made it possible to evaluate experimentally the action of spin and orbital flows. Particles are sensitive to optical forces caused both by spin and orbital flows.

Thus, the change of the measured component of the optical force leads to the change of the measurement range of: the value of the optical force associated with the spin internal energy as 10^{-25} to 10^{-10} Newton, for different types and size of particle, with the error of 10% and the value of the optical force associated with the orbital internal energy as 10^{-16} to 10^{-6} Newton with the error of 7%.

The value of the force is estimated at the pico-, femto- Newton, and in accordance to our experiments (**Figure 5**) [11] the obtained results can be considered as a proof of the mechanical action of the spin moment (spin energy flow) of the light beam on test particles of the chosen shape and properties. The experimental observation of the polarization-dependent orbital motion of test particles in a transversely inhomogeneous beam with circular polarization, where the rotational action of the orbital momentum density is absent or insignificant, is here demonstrated. Moreover, this demonstration of motion required an ultra-precise experiment, when the peculiarities of measured quantity of optical force of the femto units of Newton is taken into consideration.

In order to observe the action of optical flows on nanoparticles, an optical scheme was chosen in which the lens aperture was selected in order to avoid

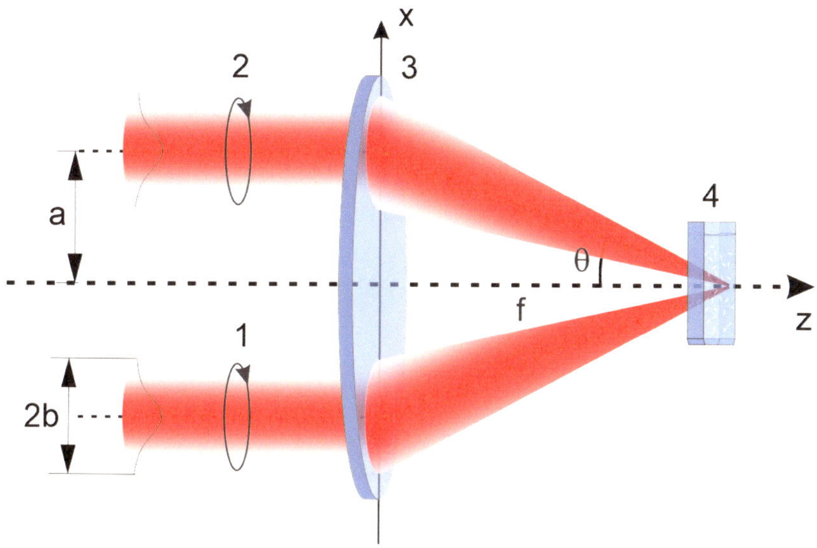

Figure 5.
Scheme of the experimental setup: (1), (2) - input beams; (3) objective lenses; (4) cuvette with test particles suspended in water.

transforming conversion of spin and orbital flows. Significant numerical apertures do not allow one to study the action of the spin momentum density, since strict focusing of a circularly polarized beam causes a partial transformation of the output spin flow into an orbital one. And even if some mechanical action of the flow is observed, then it is impossible to definitely conclude about the spin or orbital momentum action. To avoid ambiguity, focusing should not be high: the spin-orbital transformation can be neglected (i.e., it does not exceed 1%) when using lenses with a numerical aperture <0.2 (angle $\theta \approx 11°$).

Such focusing leads to a certain loss of energy concentration; however, a substantial decrease of the spin action in the focal region can be avoided if the decrease in intensity is compensated by an increase of the beam inhomogeneity. The above experimental design made it possible to reproduce and record the action of the optical force at the femto units of Newton, which is a unique confirmation of the present measurements' opportunity. The circulation of energy of a spin nature exists within each band, while the orbital momentum is completely radial and is associated with the beam divergence (**Figure 6**) [11]. This radial field momentum can be used to explain the mechanism of particle capture into the desired position outside the center, which allows one to observe the orbital motion associated with the density of the spin momentum.

In an inhomogeneous light field, any dielectric particle is exposed to the action of the gradient force that directs it to the maximum intensity, for example, to the axis of the beam, and vice versa. The radial density of the orbital momentum of the diverging beam creates a value of radial pressure that directs the particle away from the axis. As a result, both forces cancel each other out in certain areas of the interference pattern.

So, by changing the circulation of the electric field vector, the captured particle carries out orbital motion being it clockwise (**Figure 6A**) or counterclockwise (**Figure 6B**) rotational motion with respect to its own axis. Both the orbital and rotational motions stop when the polarization of the incident beam is linear. The possibility of particle transfer by the force induced by the influence of spin energy flows opens up new prospects for the creation of controlled optical micromachines,

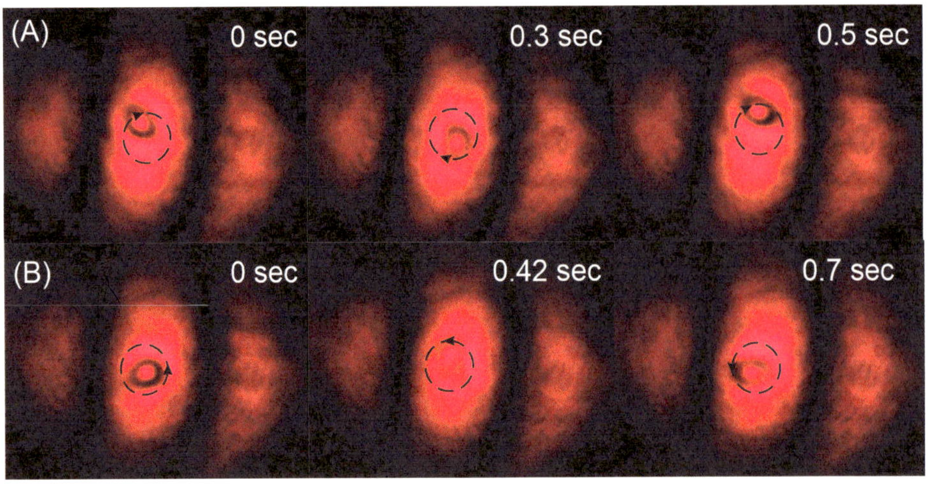

Figure 6.
The position of the captured particles within the center of the interference band at different moment of time with the change of the electric field vector circulation: A) clockwise, B) counterclockwise. The arrows indicate the direction of particle rotation.

micromanipulators in which the regulation and switching are performed through polarization control without changing the beam intensity or its spatial profile. Such methods can be useful in many systems requiring high switching speed.

It should be kept in mind that a nanoparticle is a multifunctional nanoscale tool, playing the role of a separate probe for diagnosing the field structure in the zone of a complex field microdistribution. These nanoparticles are the means of delivering microdoses of drugs into cells, and the unique means of studying complex macrostructure of the optical field in almost real time. Such a possibility can be realized in water or other transparent solutions, when the concentration of nanoparticles in various areas of the complex optical field, their spatial distribution will provide information on the spatial distribution of the amplitude, polarization, and, as a result, the phase of the field. There is an expected forecast for the study of speckle fields by nanoparticles, when they are transferred to the points of field singularity by internal energy flows, making it possible to restore information about changes of macro and micro objects over time.

2.4 Evanescent fields for micro-object manipulation

To talk about the influence of an optical field on nanostructures is necessary in order to distinguish the action of evanescent optical fields of a complex energy distribution, in particular to transfer of microdoses of drugs, providing precise accuracy of the transfer site and the transfer rate of hundreds of nanometers per second [18–20]. This situation becomes possible in the case of excitation of evanescent fields by linearly polarized beams with the azimuth of polarization of 45°. The evanescent field formed in such a way has a special distribution of spin and orbital momentum, it is elliptically polarized, in which the energy circulation is fixed in two planes - a plane parallel to the interface between the two dielectrics, where there is total internal reflection and in the plane perpendicular to it. A specific feature of the excitation of evanescent waves using a linearly polarized wave with azimuth of polarization of 45° is that the transverse spin component, which is responsible for the transverse component of the optical power appears.

In a paper [21], direct measurement of an optical momentum, which is called as extraordinary, and helicity dependent force directed perpendicularly to the

Poynting vector propagation, being proportional to the ellipticity of the local polarization of the probing beam has been reported. Such a complex structure of optical force takes place for evanescent waves and other structured fields being associated with the spin-momentum part of the Poynting vector. The extraordinary transverse momentum has been measured using a femto-newton resolution nano-cantilever immersed in an evanescent optical field above the total internal reflecting glass surface.

In our, later proposed experiments [22, 23], we use a free-standing plate for detection of the influence of an evanescent wave. We demonstrate an optically transparent birefringent microplate motion, influenced by the optical forces generated at the plate plane due to the complex optical force action: caused by the canonical momentum directed along the wavevector and the transversal spin momentum [24] directed perpendicularly to the canonical momentum. In this case it is possible to observe the motion of the plate in the direction, which does not coincide with any of these directions. The surfaces of a birefringent plate are characterized by the negligible roughness. We deposed gold nanoparticles at plate's upper surface to transfer the transverse momentum to the plate.

Elliptically polarized wave formed at a sup-surface layer enables us to estimate the longitudinal and transversal components of the force. Therefore, the vertical spin of an evanescent wave is the source of the last of them.

We simulate the spin and orbital momenta of an evanescent wave when a linearly polarized incidence wave (at second surface - boundary plate-air) with the azimuth of polarization 45° reaches the interface plate–air here undergoing total internal reflection (TIR). In this case, an evanescent wave that propagates in the z-direction, being damped in the x-direction, can be represented by [21, 24].

$$\overrightarrow{E_{ev}} = Eexp(-i\omega t)\left(\vec{x}\frac{1}{\sqrt{1+|m|^2}} + \vec{y}\frac{m}{\sqrt{1+|m|^2}}\bullet\frac{k}{k_z} + \vec{z}(-i)\frac{1}{\sqrt{1+|m|^2}}\bullet\frac{k}{k_z}\right)\bullet \exp\left(ik_z z - kx\right), \tag{1}$$

where $k_z = k\frac{n_o}{n}sin\gamma$, $\kappa = k\sqrt{\left(\frac{n_0}{n}\right)^2 sin^2(\gamma) - 1}$ is the exponential decay rate.

Here, v $m = \frac{T_\perp}{T_{//}}m_1$ is a state of polarization of an evanescent wave [24], m_1 is the state of polarization of the beam impinging at the interface plate–air being equal to unity for linear polarization with the azimuth of polarization 45°. γ is the angle of incidence at the surface, where TIR takes place.

$E = \frac{k_z}{k}\sqrt{\frac{\mu_1}{\mu}}TE_0$ is The electrical strength of the field of an evanescent wave,

where $T = \frac{\sqrt{|T_{//}|^2 + |m_1|^2|T_\perp|^2}}{\sqrt{1+|m_1|^2}}\exp\left[iargT_{//}\right]$ is the transmission coefficient [22], and $T_{//}, T_\perp$ are the Fresnel transmission coefficients.

In this case the spin momentum density has both longitudinal and transversal components [24].

Thus, the resulting momentum density in the z-direction possess both orbital and spin momentum density and is given by

$$p_z = p_{oz} + p_{sz} = \frac{A^2}{8\pi\omega}\left[\left(k_z + \frac{m^2k^2}{k_z} + \frac{\kappa^2}{k_z}\right) - 2\frac{\kappa^2}{k_z}\right]\exp\left(-2\kappa x\right). \tag{2}$$

The transversal momentum caused by the vertical spin is represented as.
$p_y = p_{sy} = \frac{A^2}{4\pi\omega}\frac{k\kappa}{k_z}\mathrm{Im}m \bullet \exp\left(-2\kappa x\right)$, where $A = E\frac{1}{\sqrt{1+|m|^2}}$.

We change the angle of incidence of a beam at the boundary plate-air. It leads to changing of the ellipticity of the evanescent wave above the plate surface. That is why one can suppose that the magnitude of the transversal spin momentum could be characterized by the different dependence towards to the resulting momentum in the longitudinal direction with the main contribution by the canonical momentum. We suppose that momentum is transmitted to the spherical surface S of the gold particles localized at the plate surface. Particle's light-scattering is taken into account within the Mie approximation [24], giving $\vec{F} = \int_S \Delta\vec{p}dS$, where $\Delta\vec{p}$ is the change of momentum density. Simulation of the force affecting a plate and causing its motion presumes integration over the illuminated area with beam aperture of 6°. The results of optical forces simulation in the y- and z-directions for a linearly polarized incident beam with the azimuth of polarization of 45° versus the incidence angle γ are shown in **Figure 7**.

As can be seen from the presented results, the transversal component of the optical force acting from the evanescent field on gold nanoparticles is experimentally fixed in our studies. Experimental study has shown that the motion of the birefrigent plate depends on the angle of incidence of the beam which generates the evanescent wave. The incidence angle influence both velocity and spinning of the plate, as well as the angle of deviation of the ahead motion from the z-axis.

Thus, our experiments [22, 23] demonstrate simultaneous plate rotation and forward plate motion. The rotation of the plate stops in the situation when the main optical axis of the plate coincides with the azimuth of polarization of the incident beam. The compensation of birefringence torques leads to vanishing of plate rotation. This is observed precisely for linearly polarized wave impinging onto the second surface of a plate at an angle ±45°. It is achieved for an angle of incidence of the probing beam about 58° which gives the azimuth of polarization of this beam at the first plate interface (boundary air-plate) equal to 62°. Therefore, rectilinear motion of the plate without its rotation is possible due to the action of the

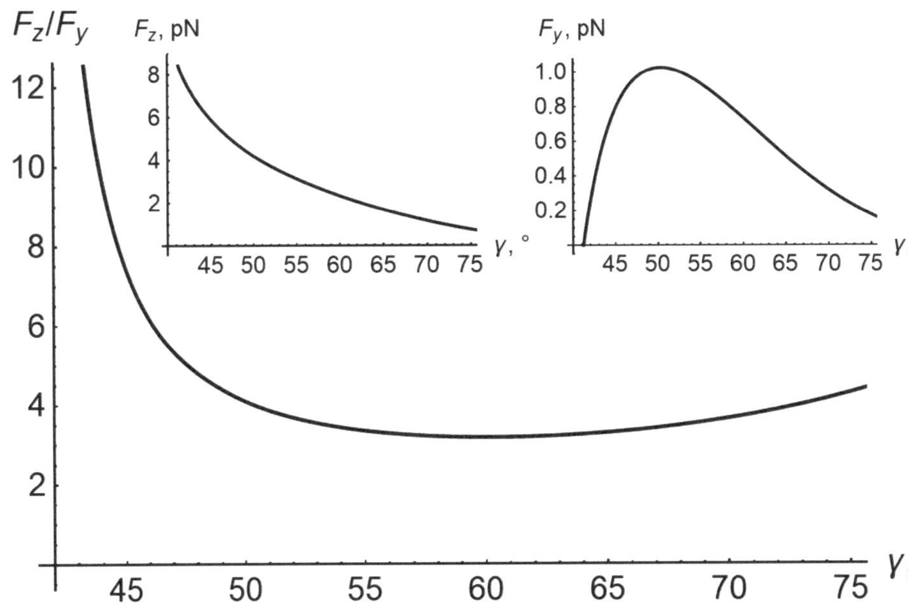

Figure 7.
The ratio of optical force in the longitudinal and transversal directions (F_z/F_y) with insets, described the resulting force in the z-direction (F_z) and the transverse diagonal polarization-dependent force induced by the vertical spin momentum in the y-direction (F_y), as a function of the incidence angle γ.

transversal component of the optical force generated above the plate. The angle between the motion direction and z-axis is 15°.

The use of these effects for evanescent waves [25, 26], in particular, is relevant for the development of nanotechnology, including molecular, especially for biomedicine, nano-therapy, transportation of medical products, bio-marking, cancer diagnosis, and bio-probing. All this provides a new tool to investigate the cell properties, i.e. mechanical or optical parameters and characteristics [26].

This brings about non-invasive methods for evaluating and analyzing pathological changes in tissues with the search for new opportunities for treating diseases and possible pathologies by non-traumatic, easily accessible methods.

2.5 Measurement of optical parameters for low-absorbing microparticles

The next step of our paper is to demonstrate one of the solutions for measurement and determining the absorption coefficient of low-absorbing microparticles by estimating the rotation speed of such objects in the field of a circularly polarized beam [27]. The uniqueness of the proposed experimental approach is that the measurement error in determining the optical parameter is of the order of femto units. It is a confirmation of the breakthrough in optical metrology and relies on the fact that modern experimental equipment and corresponding experimental approaches and measurements have gone out of the microrange to a new, more delicate level.

In particular, we present results obtained by studying the microscopic properties of liquids and various biological samples by complex optical fields [28, 29]. The optical field, acting on particles, transfers part of the momentum to them, thereby causing the spatial motion of the particles. The characteristics of this motion substantially depend on the optical constants of the studied micro- and nanoparticles.

Even the properties of complex solid samples containing, as impurities, various kinds of inclusions in the form of a fine-grained structure are determined by the optical properties of these micro and nanoparticles, thereby opening for applications. A change of the radiation propagation conditions substantially depends on the attenuation of the radiation, here determined by the absorption and scattering of radiation on these structures.

Traditionally, solutions to such problems are sought through using of Mie scattering theory, which allows one to determine mechanical ponderomotive factors (force and torque) which an optical field exerts on a particle. More importantly, the ponderomotive factors can be directly associated with the optical parameters of the particle, and this can be employed for their measurement.

In [27], the criterion of optical fields action on micro- and nano-particles is the rotational motion of the particle under the influence of a torque, which is inherent in an optical field of circular polarization with a spin angular momentum. The spinning motion of the particle is due to the field spin angular momentum absorbed by the particle, and its angular velocity Ω is related to the radiation torque by the equation $\Omega = \frac{T}{8\pi\eta r^3}$, where η = 8.9·10–3 dyn·s·cm^{-2} (at 25°C) is the dynamical viscosity of the medium. T is an absolute torque, which is calculated through Mie theory $T = 4\pi I \frac{n}{\varepsilon c}\sigma\left(Im(\alpha_e) - \frac{2k^3}{3\varepsilon}|\alpha_e|^2\right)$, here α_e – is the polarizability, appears due to the particle absorption. I is the energy flow density in the wave, c is the light velocity in vacuum, ω is the wave frequency, k = $(\omega/c)n$ is its wavenumber, and σ is the wave spin number equal to ±1 for right (left) circular polarization, respectively, and 0 for any linear or no polarization. Here the particle is assumed to be immersed in a homogeneous isotropic dielectric medium with real permittivity ε and real permeability μ so that the refraction index equals to $n = \sqrt{\varepsilon\mu}$.

The absorption index κ of the particle suspended in water and trapped in the center of a focused Gaussian beam waist with radius w_0 = 2 μm can be directly derived from the observed spinning velocity Ω exhibited by the particle in the beam with power P = 100 mW, $\kappa = q\frac{\Omega}{P}$, where q is the transition coefficient. At the same time, the particle must be with low absorption, so that there is no local heating of the medium surrounding the particle. Effective particle capture requires that the particle size be several times smaller than the size of the focusing spot, but in such a way as to prevent diffraction by the captured object. We have used weakly absorbing ($\kappa \leq 10^{-3}$) dielectric particles with diameter 0.5 to 2 μm. A schematic of the experimental equipment [27] is presented in the following **Figure 8**. This setup consists of a laser – 1, a beam expander with spatial filter –2, a mirror – 3, a quarter-wave plate – 4, objective lenses – 5, 7 and 8, cell with probing particles suspended in water – 6, CCD-camera – 9, a personal computer – 10, a control unit – 11.

As a result of the optical field action, the particle acquires a rotational motion, and the angular velocity of the particle corresponds to the part of the torque that is absorbed. As absorption increases, acceleration of rotational motion is observed. The measured value of the spinning velocity obtained in the experiment, e.g. for a gamboge particle is of about Ωe = 25.8 s^{-1}, a measurement range of the spinning velocity is 25.8 $s^{-1} \pm 0.3$ s^{-1}, the standard measurement uncertainty is 0.13 s^{-1} with the normal probability distribution. The average measured value of the spinning velocity differs from the theoretically obtained spinning velocity and determines the error of about 20%.

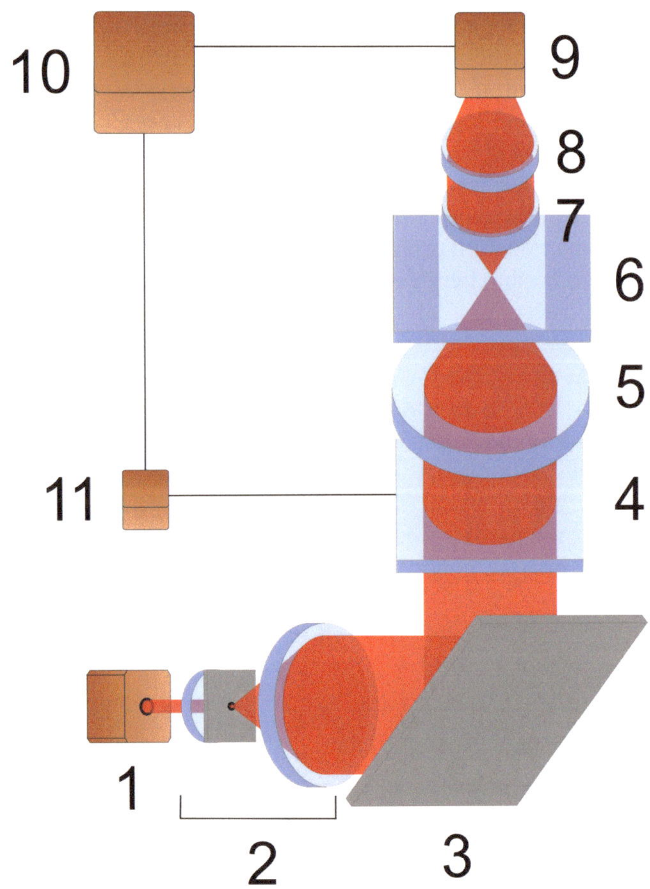

Figure 8.
Schematic of the experimental setup.

Such an error can be explained by the longitudinal displacements of the particle with respect to the beam waist, heating of the cell with particles, changes in the properties of water inside the cell, and other reasons. The introduction of the normalization coefficient, obtained by comparing the theoretical and experimental results, made it possible to determine the value of the particle absorption coefficient, here for this type of particle, it was $12.4 \bullet 10^{-4}$.

Of course, the question arises about the measurement error of the proposed method for measurement the absorption index. The flexibility of this method is determined by the refractive properties of the particle, the density of the medium where the particles are dispersed, the characteristics of the irradiating beam, and the cross-section of the focused beam. Restrictions for the determination of the particle rotation speed of about 0.1 s^{-1} are formed. Then, with the introduction of transition coefficient (q), the error in estimating the absorption coefficient δ can be obtained in the range $10^{-8} \leq \delta \leq 4 \bullet 10^{-7}$. Obtaining reliable results of measurement the absorption coefficient of particles is possible for absorption less than 10^{-2}.

Thus, the possibility of directly assessing the value of the index of light absorption by microparticles became possible thanks to the proposed approach, which uses the principles of capture and rotation of microparticles by internal energy flows. The current state of the development of technology for the manipulation of microobjects of various nature and properties makes the proposed method of measurement of optical properties encouraging and promising for many practical applications. The obtained results are the first step in developing of this kind of measurements. The high sensitivity of the absorption index measurement, the high accuracy of the estimation of the measured parameter, upon reaching the appropriate level of control for the measurement conditions, makes it possible to use this method to study weakly absorbing particles.

The above mentioned results on the involvement of the technique of micro and nanoparticles in the study of complex optical fields, the transition to control of the particle movement within a few angstroms with the acting force on the captured object at the femto Newton unit are opened the new possibilities of modern metrology. Such approaches are already used today to manipulate organelles in cells, to study viscoelastic properties, to build molecular motors and find their implementation among many other interesting applications.

2.6 Surface nanostructure and optical measurements

The next step in presenting the results of research and development on the latest methods of optical metrology is diagnostic methods for processing extremely smooth surfaces with ultra-high precision accuracy [28, 29].

Two techniques for measurement of roughness, based on measurement of a phase variance of the boundary object field and on a transverse coherence function of a field, as well as the devices implementing these techniques were proposed in our papers [28, 29]. The following principles are lying in the base of the proposed techniques:

- heights of surface microirregularities are less than wavelength of the probing radiation, (R_q < 0.1 mm), and the transversal scale of surface irregularities is larger than a wavelength, so that the specular component of the reflected radiation is present;
- phase variance is measured at the boundary field (the sample surface is imaged at the plane of analysis); the transversal coherence function of a field can be measured at arbitrary zone;

- statistical parameters of a field are measured with interferometric means, within the zero (infinitely extended) interference fringe.

- The first technique (**Figure 9**) is used for measurement of surface roughness based on measurement of a phase variance of the boundary object field [30].

Here the experimental arrangement (**Figure 9**) consists from He-Ne - laser, T - telescope, BS1, BS2- beam-splitters, O1,O2 - objective lenses, S - sample, M - mirror, PM – piezo-ceramic modulator, PD - 2x2 position-sensitive photodetector array, VC- visualization channel, EM - electric motors, AU - automatic zero fringe adjustment unit, COM - comparator, CU - analogue R_q calculation unit, DI - digital indicator.

Using an interrelation among the height parameters of surface roughness and the phase parameters of the boundary object field, one obtains the following equation for an root mean square deviation (RMS) roughness: $\boldsymbol{RMS} = \frac{\lambda}{4\pi}\sqrt{\frac{\langle I(x,y)\rangle}{I_o}}$. Technical parameters of the device are the measured RMS range – 0.002 to 0.08 μm, the measurement accuracy - 0.001 μm, the measurement scheme – micro-interferometer, indication rate - one measurement per five second, indicated units – micrometers. Here $I\left(\boldsymbol{x},\boldsymbol{y}\right), \boldsymbol{I}_0$ are the resulting field and the object beam intensity correspondingly.

This device could be used for measurement of the plane and spherical surfaces with the radius of curvature larger than 0.2 m in polishing machine tool for surface quality control during making of details. This device can be made as a stationary instrument.

The second one is a technique for measurement of surface roughness based on measurement of a field's transverse coherence function (**Figure 10**) [29].

For measurement of the degree of arbitrary surface roughness in the second technique we have proposed an experimental arrangement (**Figure 10**) with

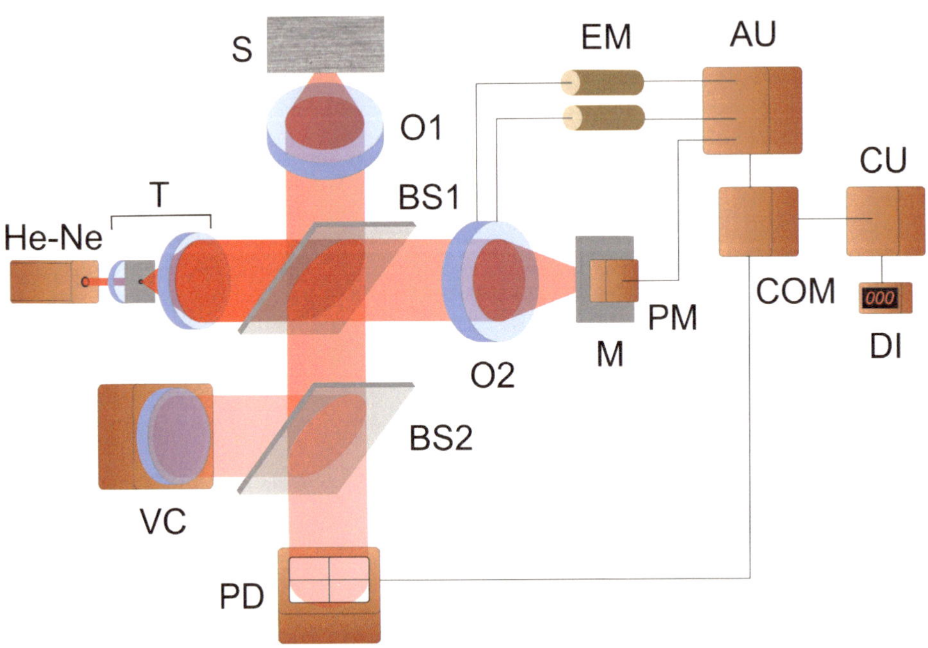

Figure 9.
Experimental arrangement for measurement of the degree of low-reflectance surface roughness.

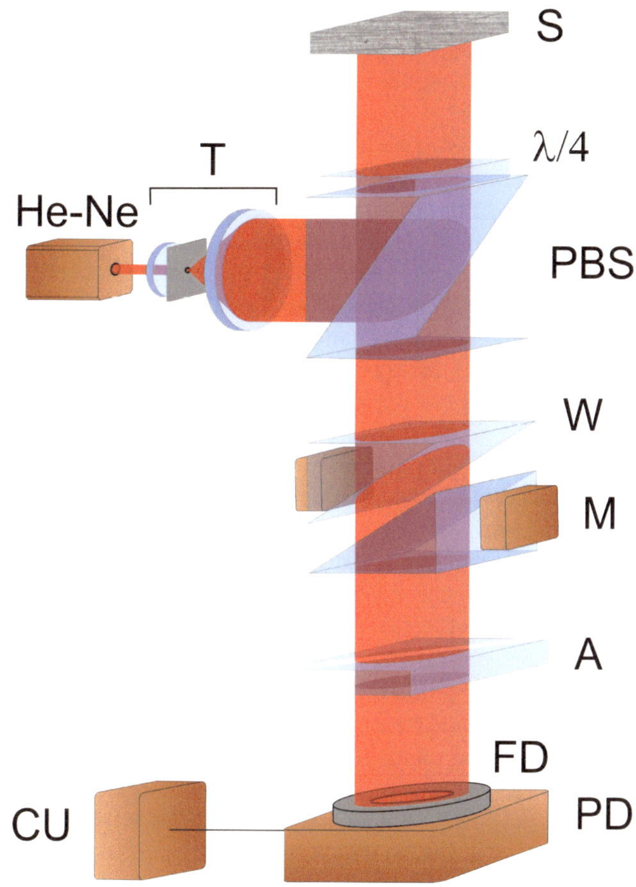

Figure 10.
Experimental arrangement for measurement of the degree of arbitrary surface roughness.

He-Ne - laser, T - telescope, PBS- polarizing beamsplitter, S - sample, W- calcite wedges; M - electromechanical modulator, A - analyzer; FD - field-of-view diaphragm; PD - photodetector, CU - analogue R_q calculation unit.

Another method for measurement of the phase variance utilizes the relationship between the transverse coherence function $\Gamma_{\perp}(\rho)$ of the scattered field, on the one hand, and the statistical parameters of the object, on the other hand. The RMS height deviation can be found from the relation $RMS = \frac{\lambda}{4\pi}\sqrt{-ln\frac{I_{max}-I_{min}}{I_{max}+I_{min}}}$, I_{max}, I_{min} are the maximum and minimum resulting intensity, respectively. The information contained in the resulting interference pattern is extracted by transforming the optical signal into electric ones with subsequent processing in the analogue electronic unit CU.

Two versions of the device for a surface roughness control based on measurement of the field's transverse coherence function were displayed:

- A stationary device that can be mounted on the processing tool. Device intended for measurement of the RMS height deviations of slightly rough surfaces over the range from 0.002 μm to 0.06 μm. Technical parameters are measured RMS range – 0.003 to 0.10 μm, measurement accuracy - 0.002 μ m. Update rate - one measurement per second. Fields of application are the following: the photochemical industry to analyze the quality of calendar shafts; arbitrarily shaped surfaces with the radius of curvature larger than 0.3 m;

polishing machine tool; space industry to monitor the quality of mirrors fabricated by diamond micro-sharpening. This device was used for surface quality on-line control.

- Portable device for control of large-area or small-area surfaces. Portable device intended for measurement of the RMS height deviations of slightly rough surfaces. Main technical parameters of the device are the measured RMS range – 0.005 to 0.10 μm, the measurement accuracy - 0.002 μm, the update rate - one measurement per second. This device provides the following field of application: device can be made either as a measuring head, or as a stationary instrument, depending on the size and the position of the object to be controlled; polishing machine tool, this device was used for the surface quality control during making of the detail;.

Sensitivity of the RMS height parameter of all these devices down to 10 Å was achieved. Roughness control of slightly rough (polished) surfaces with RMS deviation of a profile from a mean surface line ranging from 0.002 μm to 0.10 μm. The technique is applicable to metallic, insulator, semiconductor, and optical surfaces.

2.7 Optical refractive index measurement

At last it would like to bring one more aspect of fine measurement related to the determination of such an optical parameter of the studied object as the refractive index of light-scattering liquid media. Here, a specific issue is polarization interferometry to find the refractive index of solutions, suspensions and gaseous media [31]. According to the proposed approach [31], a two-beam interferometer is used to determine the refractive index, in which a circularly polarized beam is formed in each channel. The measurement method consists in splitting the optical radiation into two components and forming mutually orthogonal circular polarizations of the field components. As a result of superposition of such beams, a uniform intensity distribution is seen in the interference plane. In this case, the formed field has some deterministic polarization. The resulting field is linearly polarized and is characterized by the azimuth of polarization α_0. When the studied solution is located in the object channel giving rise to polarization change of the transmitted radiation.

As a result, a rotation of the azimuth of linear polarization is observed in the interference plane, which takes on the value α. That is, the difference in polarization azimuth $\alpha - \alpha_0$, which is related to the reduced path difference of the beams in the arms of the interferometer, and is given by $n = n_0 + \Delta n_0 = n_0 + \frac{(\alpha-\alpha_0)\lambda}{2\pi l_0}$, where n_0 is the refractive index of the main medium (solvent), l_0 is the cell thickness.

The azimuth of polarization of the resulting distribution is sensitive to variations in the path difference in the arms of the interferometer. Even with a change in the path difference between the orthogonal components by an amount less than λ, the polarization azimuth will change significantly. If the path difference between the orthogonal components is λ, then the polarization azimuth will change by 2π. The polarization azimuth can be measured with a measurement error of some seconds. As a result, the accuracy of path difference measurement (Δl) is $10^{-5}\lambda$, and the measurement accuracy of the value of the refractive index is $10^{-6}\frac{\lambda}{l_0}$.

2.8 Biomedical application of 3D laser polarization metrology

Now we will try to present some last results on the development and experimental testing of a new 3D Stokes-polarimetry method for mapping the object fields of biological optically anisotropic layers [8–10]. Particular, we propose digital holographic layer-by-layer reconstruction of polarization elliptisity distributions to express diagnostic and differentiate of polycrystalline blood films samples of patients with prostate tumors. The structural-logical diagram and design of 3D layer-by-layer Stokes-polarimetry method are described by **Table 1**.

The proposed optical scheme is shown in **Figure 11**.

Collimator 2 forms a parallel ($\text{Ø} = 2 \times 10^3$ μm) beam of He-Ne ($\lambda = 0.6328$ μm) laser 1, which is divided by 50% beam splitter 3 into "irradiating" (Ir) and "reference" (Re) ones. The "irradiating" beam is directed through the polarizing filter 6–8 by the rotating mirror 4 on the sample of the biological layer 9. The polarization-inhomogeneous image of the object 9 is projected by the lens 10 into the plane of the digital camera 15. The "reference" beam is forwarded by the mirror

1	**Optical probing source**	**Gas helium-neon laser; Wavelength 0.6328 μm**
2	Block for forming the spatial structure of the optical probe	Optical collimator for forming a parallel laser beam with a cross section of 5 mm
3	Block for the formation of the polarization structure of the optical probe	System of formation of linear (0°, 90°, 45°) and right-circular polarization (linear polarizer.
4	Object block	Microscopic coordinate node
5	Formation block of microscopic image	Polarizing micro lens
6	Multichannel polarization filtering block	Transmission system of linearly (0°, 90°, 45°, 135°), right- and left-circularly polarized components
7	Reference wave formation block	Polarizing beam splitter
8	Block for the formation of the polarization structure of the reference coherent wave	Linear (0°, 90°, 45°) and right-circular polarization system
9	Polarization filtering block	Linear polarizer
10	Discritization block	Digital CCD camera
11	Block for computer processing of polarization interferometry data	Calculation algorithms: • distributions of the magnitude of the ellipticity • statistical moments of the 1st - 4th orders.

Table 1.
Structural-logical diagram of the method of 3D layer-by-layer stokes-polarimetry.

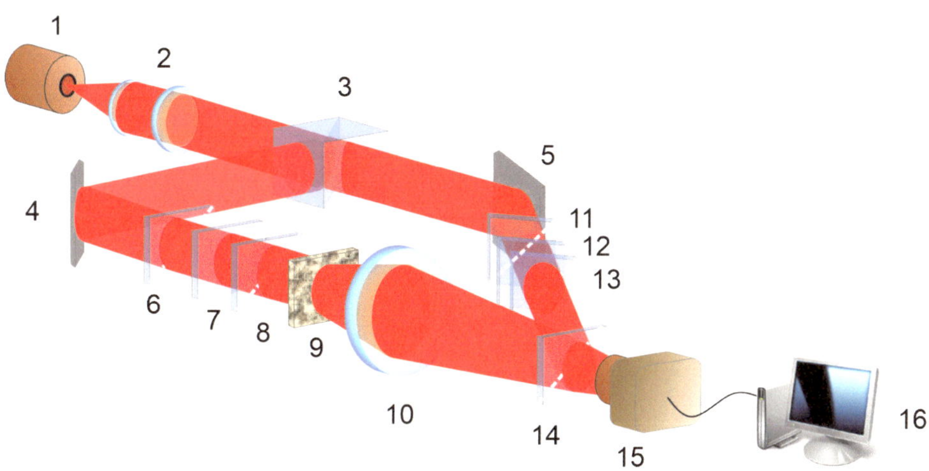

Figure 11.
Optical scheme of 3D Stokes-polarimetry for polycrystalline blood films microscopic images.

5 through polarizing filter 11–13 into the object image plane 9. The resulting interference pattern is recorded by a digital camera 15 through the polarizer 14.

As a result, the topographic maps and histograms of statistical parameters and polarization ellipticity of layer-by-layer digital microscopic images of polycrystalline blood films of patients were measured and compared. It made possible to carry out a differential diagnostic of benign and malignant prostate tumors with different degrees of differentiation.

3. Conclusions

The results presented in this chapter provide an overview of the new approaches of measurements and results obtained by the authors of this chapter over the past decades, in the direction of ultra-sensitive precision measurements. In this chapter we have tried to present the metrological aspects of new optical measurements, that had made it possible to expand the range of measurement of such optical quantities as optical force and optical parameters of solid, liquid and gas up to femto and pico units. This improves the accuracy of optical measurements for the corresponding quantities by 3 orders of magnitude.

The sensitivity of the methods achieved by interference, polarization-interference when assessing the roughness of smooth surfaces, when measurement variations in the refractive index of aqueous solutions ranges from tens of angstroms to tens of nanometers.

The femto unit of optical forces, which occurs in complex optical fields with a rich morphology of the distribution of internal optical energy flows, interacts with micro- and nano-objects of various shapes and properties, but also control their spatial motion in optical traps, the nature of which can differ significantly in accordance with the trap formation mechanism. The optical forces of this physical nature and the values are used to determine the optical parameters of trapped particles, where the error of the determination is controlled at the nanoscale level.

The nature of the trap is different, but the original and new of our approaches are the use of biaxial crystals, which allowed us to form a complex distribution of internal energy flows and at the same time realize the conversion from phase-amplitude to polarization structures intended for nano-objects capture.

New polarimetric solutions for 3D Stokes mapping of microscopic images of polycrystalline blood films with digital holographic reproduction of layer-by-layer polarization maps of ellipticity with the success is shortly demonstrated in the given chapter.

It is clear that this review cannot be a complete analysis of new modern ultrathin methods of optical metrology. It is also worth recalling the ideas and principles of STED microscopy [32, 33], the physical principles of the operation of optical vortex coronographs [34], and much more. As an exclusive it could be noted the fundamentally achievable opportunity to operate and manipulate with atoms [2] - this is the ultimate fantasy for the optical range.

DOI: http://dx.doi.org/10.5772/intechopen.100589

Author details

Oleg Angelsky[1,2], Peter Maksymyak[2], Claudia Zenkova[1,2], Olexander Ushenko[2] and Jun Zheng[1*]

1 Research Institute of Zhejiang University, Taizhou, China

2 Chernivtsi National University, Chernivtsi, Ukraine

*Address all correspondence to: dbzj@netease.com

References

[1] Hell S, Nanoscopy with focused light (Nobel Lecture), Angew. Chem. Int. Ed. Review. 2015; 54:8054–8066.

[2] Rubinsztein-Dunlop H. et al., Roadmap on structured light. J. Opt. 2017; 19:013001.

[3] Rotenberg N, Kuipers L, Mapping nanoscale light fields. Nature Photon. 2014; 8: 919.

[4] Bauer T, et al., Observation of optical polarization Möbius strips. Science. 2015; 347: 964.

[5] Geng J, Structured-light 3D surface imaging: a tutorial. Adv. Opt. Photon. 2011; 3: 128–133.

[6] Ghosh N, Tissue polarimetry: concepts, challenges, applications, and outlook. J. Biomed. Opt. 2011; 16: 110801.

[7] Vitkin A, Ghosh N, & de Martino A, Tissue Polarimetry. In: Photonics: Scientific Foundations, Technology and Applications. Andrews DL, editors. John Wiley & Sons, Ltd.; 2015, p. 239–321.

[8] Angelsky OV, Ushenko A, Ushenko YA, Pishak VP & Peresunko AP. Statistical, Correlation, and Topological Approaches in Diagnostics of the Structure and Physiological State of Birefringent Biological Tissues. In Handbook of Photonics for Biomedical Science. Tuchin V V. editor. CRC PRESS; 2010, p. 283–322. doi:10.1201/9781439806296-c10

[9] Ushenko VA, Hogan BT, Dubolazov A, Grechina AV, Boronikhina TV, Gorsky M, Ushenko AG, Ushenko YO, Bykov A, Meglinski I, Embossed topographic depolarisation maps of biological tissues with different morphological structures, Scientific Reports 2021; 11(1): 3871.

[10] Peyvasteh M, Tryfonyuk L, Ushenko V, Syvokorovskay A-V, Dubolazov A, Vanchulyak O, Ushenko A, Ushenko Y, Gorsky M, Sidor M, Tomka Y, Soltys I, Bykov A, Meglinski I, 3D Mueller-matrix-based azimuthal invariant tomography of polycrystalline structure within benign and malignant soft-tissue tumors. Laser Physics Letters 2020; 17 (11): 115606.

[11] Angelsky OV, Bekshaev AYa, Maksimyak PP, Maksimyak AP, Hanson SG, and Zenkova CYu, Orbital rotation without orbital angular momentum: mechanical action of the spin part of the internal energy flow in light beams. Opt. Express 2012; 20(4): 3563–3571.

[12] Angelsky OV, Bekshaev AYa, Maksimyak PP, Maksimyak A., Mokhun II, Hanson SG, Zenkova CYu, and Tyurin AV, Circular motion of particles suspended in a Gaussian beam with circular polarization validates the spin part of the internal energy flow. Opt. Express 2012; 20(10): 11351–11356.

[13] Bekshaev AYa, Angelsky OV, Hanson SG, and Zenkova CYu, Scattering of inhomogeneous circularly polarized optical field and mechanical manifestation of the internal energy flows. Phys. Rev. A 2012; 86(2): 023847.

[14] Zenkova C, Soltys I, Angelsky P, The use of motion peculiarities of particles of the Rayleigh light scattering mechanism for defining the coherence properties of optical fields. Optica Applicata 2013; 43(2): 297–312.

[15] Angelsky OV, Bekshaev AYa, Maksimyak PP, Polyanskii PV, Internal Energy Flows and Optical Trapping. Optics & Photonics News (OPN) 2014; 25.

[16] Dienerowitz M, Mazilu M, Dholakia K, Optical manipulation of

nanoparticles: a review. J. Nanophoton 2008; 2: 021875.

[17] Bekshaev AYa, Subwavelength particles in an inhomogeneous light field: Optical forces associated with the spin and orbital energy flows. J. Opt. 2013; 15: 044004.

[18] Angelsky OV, Maksymyak PP, Zenkova CYu, Maksymyak AP, Hanson SG, Ivanskyi DI, Peculiarities of control of erythrocytes moving in an evanescent field. J. of Biomedical Optics 2019; 24 (5): 055002.

[19] Angelsky OV, Zenkova CYu, Ivanskyi DI, Mechanical action of the transverse spin momentum of an evanescent wave on gold nanoparticles in biological objects media. Journal of Optoelectronics and Advanced Materials 2018; 20(5–6): 217–226.

[20] Angelsky OV, Zenkova CY, Maksymyak PP, Maksymyak AP, Ivanskyi DI, Tkachuk VM, Peculiarities of Energy Circulation in Evanescent Field. Application for Red Blood Cells. Optical Memory and Neural Networks (Information Optics) 2019; 28(1): 11–20.

[21] Antognozzi M, Bermingham CR, Hoerber H, Dennis MR, Bekshaev AYa, Harniman RL, Simpson S, Senior J, Bliokh KY, Nori F, Direct measurements of the extraordinary optical momentum and transverse spin-dependent force using a nano-cantilever. Nature Physics 2016; 12: 731–735.

[22] Angelsky OV, Hanson SG, Maksimyak PP, Maksimyak AP, Zenkova CYu, Polyanskii PV, and Ivanskyi DI, Influence of evanescent wave on birefringent microplates. Optics Express 2017; 25(3): 2299.

[23] Zenkova CYu, Ivanskyi DI, Kiyashchuk TV, Optical torques and forces in birefringent microplate. Optica Applicata 2017; 47(3): 483–493.

[24] Bliokh KY, Bekshaev AY, Nori F, Extraordinary momentum and spin in evanescent waves, Nature Communications 2014; 5.

[25] Yoon Y-Z, and Cicuta P, Optical trapping of colloidal particles and cells by focused evanescent fields using conical lenses. Opt. Express 2010; 18(7): 7076.

[26] Gu M, Kuriakose S, Gan X, A single beam near-field laser trap for optical stretching, folding and rotation of erythrocytes. Opt. Express 2007; 15(3): 1369.

[27] Angelsky OV, Bekshaev AYa, Maksimyak PP, Maksimyak AP, and Hanson SG, Measurement of small light absorption in microparticles by means of optically induced rotation. Opt. Express 2015; 23(6): 7152–7163.

[28] Angelsky OV, Maksimyak PP, Hanson S. The Use of Optical-Correlation Techniques for Characterizing Scattering Object and Media. SPIE Press PM71 Bellingham, USA; 1999.

[29] Angelsky OV, Maksimyak PP. Optical correlation diagnostics of surface roughness" in Optical Correlation Applications and Techniques. SPIE Press, Bellingham; 2007.

[30] Angelsky OV, Maksimyak PP. Optical correlation diagnostics of surface roughness in coherent-domain optical methods. In: Tuchin VV, editor. Biomedical Diagnostics Environmental and Material Science. Kluwer Academic Publishers; 2004.

[31] Angelsky OV, Maksymyak PP, Polyansky VK, Measurement of refractive index of light scattering media (Avtorskoye svidetelstvo). 1986 (in Russian).

[32] Punge A, Rizzoli SO, Jahn R, Wildanger J D, Meyer L, Schonle A,

Kastrup L, Hell SW, 3D reconstruction of high-resolution STED microscope images. Microscopy Research and Technique 2008; 71: 644–650.

[33] Vicidomini G, Bianchini P, Diaspro A, STED super-resolved microscopy. Nat. Methods 2018; 15: 173–182.

[34] Valle PJ, Fuentes A, Canales VF, Cagigas MA, Villo-Perez I, and Cagigal MP, Digital coronagraph algorithm. OSA Continuum 2018; 1(2): 625–633.

Chapter 4

Calibration of Tanks and Ships' Tanks for Storage and Transportation of Liquids by Laser Scanning

Oleksandr Samoilenko and Volodymyr Zaets

Abstract

The goal of the research is to improve the accuracy of measurement the volume and mass of oil and oil products by the stationary measuring tanks and ships' tanks. It is possible to achieving this goal only by using the laser scanning at tanks calibration. Metrological and other technical requirements for laser scanners have been developed. It is proved by the results of mathematical modeling that only the compliance of scanners with the developed requirements makes it possible to achieve the set goal. It has been developed methods of measurements by laser scanners that allow to achieve an increase in the accuracy of determination the interval capacities of all types of the tanks. Methods, formulae and algorithms for interval capacities of tanks calculation are very complicated. Therefore, the interlaboratory comparisons for the interval capacities, calculated by laboratories own software developed for processing the results of the specified calibration are proposed. It is concluded that the developed requirements, methods and algorithms will allow, in several times, to increase the accuracy of determining the interval capacities of the tanks with a significant reduction in time for measurements and processing of their results.

Keywords: liquids volume and mass measurement, laser scanner, vertical and horizontal cylindrical tanks, spherical tanks, ship tanks, calibration, measurement uncertainty

1. Introduction

High speed laser scanners (to 2 million points per second) in automatic mode measure horizontal and vertical angles and distances to any visible point of the tank in non-reflective (non-contact) mode. According to the measurement results, the coordinates x_i, y_i, z_i of the points on the surface with a standard uncertainty of 2–3 mm are immediately calculated and entered into a file in the device's memory. Many laboratories apply manual electronic and scanning total stations, but in speed of measurement, they significantly yield to scanners.

Measurements can be performed at a wide range of temperatures and in strong winds with modern scanning devices, and with some types of instruments, even in the rain.

The scanner scans everything that is around within its line of sight except for a small "dead" zone under the device, and when working "upside down", above the device.

One of the ways to increase accuracy of volume and mass measurement of liquids stored in vertical and horizontal cylindrical, and spherical tanks and ships' tanks is a laser scanning of their surface in the process of their calibration. This significantly increases the productivity of the measurement process. Without special software it is impossible to fulfill rather complicated calculations for accurate estimation of the interval capacities of tank and their uncertainties. Correct use of the scanners for coordinate measuring, preparation of tank's 3D model, development and use of special software for tank capacity calculation and its uncertainty calculation is impossible without clear definition of those processes and algorithms. The crude errors and unproven methodic of laser scanning cause crude errors in measurement of a tanks' capacity up to 1–5%. The nonconformity of laser scanners to set requirements causes errors in measurement of the tank's capacity up to 0,5–2%.

2. The main goal and tasks of the research

The main goal of these research are the development of the laser scanners measurement methods and calculations of the tank's interval capacities methods, which provide an increase in the accuracy of measuring the capacity of tanks in relation to the requirements of OIML R71 [1].

Simultaneously, the task of shortening the measurement time using laser scanners and shortening the processing time and creation of a tank capacity table based on laser scanning data using special software is being solved.

The internal capacities of the tank are entered into the capacity table. It sets up a correspondence between absolute height of the level of liquid in the tank and its capacity. So, the interval capacity is the value of the tank capacity on the pointed out in the capacity table absolute height of the liquid level.

Many methodical details concerning correct use of the laser scanners for tanks calibration exist. These details, as a rule, are absent in the valid standards that provide application electro-optic method for measurement at tanks' calibration [2–10]. The standards [11–13] can be considered the most complete but they are dated and need to be revised taking into account the latest calibration experience.

This publication is aimed to present clear formulation of important technical requirements concerning the correct usage of scanners in order to ensure fulfillment of the main goal. According to the existing data crude errors and unproven methodic of laser scanning cause crude errors in measurement of tanks capacity up to 1–5%.

Not all types of the laser scanners are appropriate for measurements in the process of tanks' calibration or they have certain constraints for calibration of particular tanks types. Ensuring fulfillment of the main goal should include clear metrological and technical requirements to the laser scanners. Laboratory experience proves that nonconformity of the laser scanners to set requirements causes errors in the measurement of the tank's capacity up to 0,5–2%.

Shortening of the measurement time, i.e., fulfillment of the task of shortening the measurement time is provided almost by any scanner, but its use is appropriate only in case, if it conforms to the set requirements and is used correctly for appropriate tank types.

Previous processing of number of points, which coordinates are measured by laser scanner on the surface of the tank walls (files with approximate size from 50 to 200 MB) has a great importance for fulfillment of the main goal and tasks. The result of such previous processing is a 3D tank model, applicable for further calculation of its interval capacities and their uncertainties. This procedure and requirements to it should also be clearly formulated.

Methods, formulae and algorithms for calculation interval capacities of tanks are very complicated. Fulfillment of the main goal depends on accuracy of methods, formulae and algorithms for calculation of interval capacities of tank. Moreover, they should provide calculation for tanks scanned internally and externally, for all complicated cases of spatial position, for example, huge tank axis tilt, big deformations of tank walls and structural features of tank, etc. To fulfill the main goal and of the task of shortening the processing time and creation of a tank capacity table, it is necessary that they should be clearly and definitely formulated in the new standards. This will help to develop special software and test it for fulfillment of the main goal and decision the tasks.

For final fulfillment of the task of shortening the processing time and creation of a tank capacity table, it is necessary that special software automatically formulate calibration certificate including tank capacity table with interval capacity uncertainties and other important data. Form of the measurement report and certificate should be clearly formulated.

To fulfill the task of shortening the processing time and creation of a tank capacity table, it is necessary to develop special software. Calibrating laboratories have no economical or technical possibilities to develop the mentioned software. There are examples of software development for processing calibration results of tanks by laser scanners manufacturers. Thus, having no experience in calibration of tanks and clear requirements, these attempts cannot be called successful [14].

Fulfillment of the main goal will allow to provide:

- accurate bills calculations for shipped liquids in the process of commercial and tax operations, including fraud protection;
- accuracy increase of internal accounting of liquids quantity on the enterprises and organizations, which should shorten nonproductive loses of liquids and improve internal financial reporting indices;
- accuracy increase of liquids batching in the process of technological operations of mixing or preparing solutions, etc.

Fulfillment of the task of shortening the measurement time will allow to:

- save by enterprises and organizations owing to substantial shortening of nonproductive using of tanks and connected to it technical equipment. For example, calibration of 4–5 tanks on the filling station by laser scanning method taking 40–60 minutes, while using volumetric method according to ISO 4269:2001 [15] takes from 1 to 3 days.

Fulfillment of the task of shortening the processing time and creation of a tank capacity table will allow:

- organizations concerned to develop software for processing calibration results more applicable for fulfillment of the main goal;

- calibration laboratories to get software, which will shorten processing of measurement results time and will form calibration certificate of tanks and their capacities tables;
- accuracy increase of liquid volume measurement in the process of commercial, tax and technical operations, also internal accounting of liquids circulation, for example, oil and gas.
- complete cancelation of calibration of tanks using ISO 4269: This will help to save on large volume of dirty water filtering, which was used for tank calibration. Besides, accuracy of capacity measurement will remain the same or increase.
- clear recommendations will allow to shorten substantially crude and systematic errors quantity in the process of tanks calibration and calculation of their capacity by calibrating laboratories.

3. Sources analyses

One of the obstacles to achieve goal set in the Section 1 is dated standards [2–13]. Standards GOST P 8.994 [7] and GOST P 8.996 [8] are new and they provide usage of total stations and scanners but schemas, measurement methods and capacity calculation methods as in [2–6, 9, 10] do not meet the above set goal.

OIML R 71 [1] does not contain requirements to field measurement methods, tank capacity calculations and software. In the tank calibration it refers to the international standards [2–5].

In the standards [2–10] the tank is not considered as single spatially closed surface of a geometrically regular or free form where spatial coordinates of each point are known in a single spatial coordinate system.

According to the standards [2–10] the radius of the cylindrical tank is calculated from coordinates approximation in separate unconnected plane sections. As a result, part of the significant information about shape and size of the tank is not used. It causes latent (almost impossible to estimate) increase in the uncertainty of the interval capacities calculation. Determination of the axis tilt of a vertical tank to a plumb line or of a horizontal tank to horizon in these standards is performed by some separate measurements. Tilt definition is not associated with mentioned sections. The axis tilt uncertainty is unsatisfactory. In [12, 13, 16] the axis tilt and its uncertainty are estimated strictly by the least square method (LSM) for all points of the 3D model.

To achieve the goal spatial coordinates must be used to calculate parameters and capacities of the tanks. This is implemented in the standards [11–13], as well as [16, 17]. Though these standards also require substantial revision taking into account the experience received in recent years.

All listed standards do not contain sufficient information as to requirements to optical-electronic devices for tanks' calibration. A minimum of metrological characteristics and parameters of devices is listed.

So called verification of total stations, described in ISO 7507-4 [3] and ISO 12917-2 [5], is incorrect and harmful procedure for determination suitability of the devices for tank calibration. It is necessary to calibrate scanners with setting specific metrological characteristics and parameters that affect the uncertainty of capacity calculation by point coordinates on the tank walls surface.

For many years, researches have been carried out on metrological and other technical characteristics and parameters of scanners. However, the research of their influence on uncertainty in calculating tank' interval capacities has not been carried out. Therefore, references on multiple sources regarding scanners research are not provided.

In the standards ISO 12917-2 [5], GOST P 8.994 [7] and GOST 8.659 [13] it is proposed to use total stations for calibration horizontal cylindrical tanks. Many years of experience show that the use of the total stations has no prospect for the calibration of such tanks. Most of them are small in size. The diameter ranges from 1 to 3,5 m. The total stations have range in measurement distances in non-reflective mode 1,2–1,5 m what makes their usage very difficult in tight working space inside the tank. Labor productivity at calibration of such tanks by total station according to ISO 12917-2 [5], GOST P 8.994 [7] and GOST 8.659 [13] does not stand up to criticism. It can last 3–8 hours versus 10–15 minutes required for scanning the tank from inside and 15–30 minutes from outside. Using a total station during 3–8 hours the number of points on the tank wall surface is not enough to calculate interval capacities with uncertainty better than 0,5–1%.

Attempts to apply method of capacity calculation by areas of separate horizontal sections to spherical tanks [18] have no prospects at all. Only in GOST 8.655 [11, 17] mathematically strict approach is used when an approximation by spatial geometrical figure is applied to coordinates of all point on the surface. Then, the capacity of the spatial figure segment is calculated below the specified horizontal plane. A correction for the relief of the real surface is added relative to the approximated one with the using all points of the segment. In this case uncertainty is estimated mathematically strict. A similar approach is used in GOST 8.656 [12], GOST 8.659 [13, 16].

A special place among scanners is occupied by highly specialized scanning systems [19, 20]. Their clear advantage is in explosion-proof design. It is very attractive for tank owners that no cleaning and degassing of the tank is required. Scanning is carried out through a relatively small hole in the tank hatch. Duration of scanning is 2 hours [20]. But apart from small horizontal tanks from the inside, they cannot scan anything else. There are big doubts about the declared accuracy [20] (the declared accuracy is absent in [19]). The reasons for these doubts will be stated below. Scanner calibration certificates and any materials of comparative tests of the described technology stated in [19, 20] are not accompanied.

4. Requirements to devices for measuring coordinates on the tank walls

4.1 The main geometrical conditions of the devices for measuring coordinates on the tank walls

Let us briefly recall some important geometrical conditions that must be met for any scanner or total station used for tank calibration.

Rotation axis of the device must be vertical. It is often called the vertical rotation axis, although it is never exactly vertical. The device must measure tilt of the rotation axis of the device and transform the measurement results and calculated coordinates so that axis z of the spatial coordinate system is vertical and the plane xy is horizontal.

Rotation axis of the scanner mirror or the telescope of the total station must be perpendicular to rotation axis of the device. It is often called horizontal rotation axis although it is only approximately horizontal. The point of intersection of the rotation axes of the device and the mirror or tube can be called the reference point of the device, since the zero of the system of measured spatial coordinates is placed exactly in it.

The laser beam axis of the scanner distance meter, which is directed to the tank for measuring the distance, must be perpendicular to the axis of rotation of the mirror or tube. The distance to the tank wall should be measured from the reference point of the device. In reality, it is measured from some other point located on the laser beam axis. The distance from this point to the reference point of the device is called the additive constant of the distance meter of the device [21, 22].

All devices measure cylindrical coordinates – horizontal and vertical angles and slope distance α, v, D with subsequent transformation into a spatial rectangular coordinate system. The zero of these systems is located at the reference point of the device.

4.2 Requirements to scanners depending on the capacity of the calibrated tank

If calibration laboratory has scanners with the metrological characteristics and parameters that are recommended in **Table 1** it does not guarantee the achievement of the set tasks for reducing the uncertainty of interval capacities. However, the lack of scanners with such metrological characteristics and parameters makes achievement of the main goal very difficult.

The expansion coefficient is equal to 2 for the expanded uncertainty in **Table 1** at a confidence level of 0.95 and a normal distribution law.

	The name of the metrological characteristic or parameter of the scanner	**Nominal capacity of the tanks, m^3**			
		Up to 50	**From 50 to 1000**	**From 1000 to 5000**	**Above 5000**
1	Approximate range of measured slope distances to the tank wall, m	0,6–6	1–10	2–20	3–50
2	Standard deviation of the coordinates' measurement in the range of the measured distances (random component of the measurement), mm, not more	1,5	2,2	3	4
3	Additive constant of the scanner distance meter in the range of measurement, mm, not more	5	10	15	20
4	Expanded measurement uncertainty of the additive constant of the scanner distance meter at its calibration, mm, not more	0,5	1,0	1,5	2
5	Measuring range of the tilt of the scanner rotation axis when working in the inverted state, ... °, not less	3	3	3	3
6	Expanded uncertainty of measuring the tilt of its rotation axis by the scanner, ... ′, not more	1	1	1,5	2
7	Index error (zenith point), ... ′, not more	1	1	1,5	2
8	Collimation error – deviation from the perpendicularity of the laser beam axis and mirror rotation, ... ′, not more	2	2	2	2
9	Deviation from the perpendicularity of mirror rotation axis and the laser scanner rotation axis, ... ′, not more	2	2	2	2
10	Expanded uncertainty of measuring the absolute height of the scanner reference point relative to the tank reference point, mm, not more	2	3	4	5

Table 1.
Recommended metrological characteristics and parameters of scanners depending on capacities of the calibrated tanks.

Some comments to items of **Table 1**.

1. In the first line of **Table 1** it is given the most probable range of distances in which the scanner has to work at calibrating a tank of the specified capacity. It should be noted that not all types of scanners provide the minimum measurement range.

2. If you scan a smooth surface, at the scan it will not look smooth, but will have some roughness. It is characterized by a standard deviation.

3. The additive constant of the scanner distance meter should be zero, but this is never the case. It is recommended that it does not exceed the values given in **Table 1**. Unfortunately, it can have slightly different values at different sub-ranges of the measured distances. This should also be investigated and taken into account when introducing amendments.

4. Special attention should be paid to the measurement of the additive constant of the distance meter, as it should be used to introduce corrections to each point coordinate on the surface of the tank walls. Its uncertainty directly affects the uncertainty in determining the size of the tank, and, consequently, the uncertainty in determining its capacity. For example, it is highly desirable that for horizontal tanks with a nominal capacity of 5 to 50 m^3, the expanded uncertainty of the mean radius does not exceed 0,2–0,5 mm, and the average length does not exceed 1–2 mm.

5. It is difficult to position quickly and accurately the rotation axis of the scanner in a vertical position when it is upside down. Therefore, it is very important to have a wide measurement range of the tilt of the scanner's rotation axis in an inverted state (**Figure 1**). If the scanner is used only in a normal vertical position, then the measurement range of the rotation axis tilt can be $\pm$ (5′ – 15′). In this case, the rotation axis tilt of the scanner inside vertical tanks can be significantly affected by the instability of its bottom.

6. Uncertainty in measuring the tilt of the scanner rotation axis is a very important characteristic, since the random and systematic components of the tilt measurement are directly contributed into the uncertainty of the axis tilt of the cylindrical tank.

7. Theoretically, the zero of the devices for measuring vertical angles should coincide with the rotation axis of the device, but this is not always the case. The angle between the rotation axis, which must be vertical and the zero of the devices, is called the index error.

8. Collimation error – deviation from the perpendicularity of the laser beam axis and tube rotation for total station should not exceed 0,5′.

9. Deviation from the perpendicularity of tube rotation axis and the total stations rotation axis is should not exceed 0,5′.

10. This item refers to the use of the scanner at tank calibration. The absolute height of the scanner reference point relative to the tank reference point should be determined before and/or immediately after scanning the tank with the recommended uncertainty.

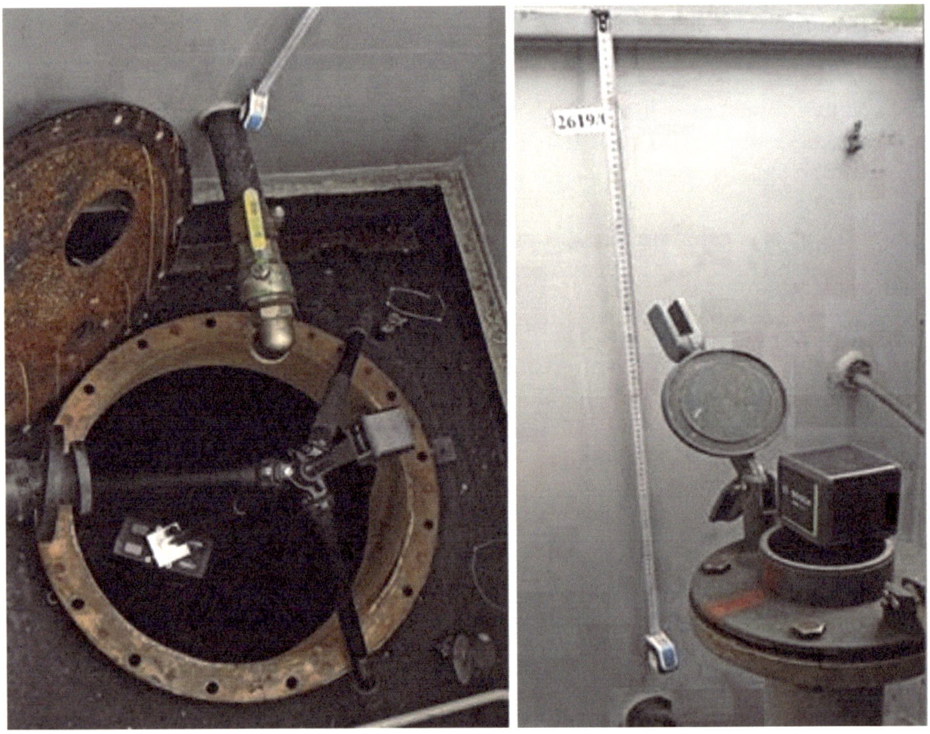

Figure 1.
Scanning of the horizontal cylindrical underground tank through its hatch.

Based on the above information, the scanners described in [19, 20] should be subjected to rigorous metrological testing. For them, the requirements of items 2, 4, 6–10 must be met. Otherwise, it is impossible to talk about the achievement of the set goal regarding the uncertainty of interval capacities less than 0,2%. It should be noted that some requirements for such a design of scanners should be formulated somewhat differently.

4.3 Modeling the influence of systematic measurement errors and deviations of the device's parameters from the nominal values on the calculated interval capacities

To demonstrate the influence of systematic measurement errors and deviations of the device's parameters from the nominal values on the calculated interval capacities, a simulation was performed when the coordinates of points on the tank surface were deliberately distorted by errors and deviations.

A horizontal cylindrical tank with flat bottoms with a nominal capacity of 29 m^3 was chosen as a model (**Figure 2**). Its length is approximately 4,78 m, and its diameter is 2,78 m. The prepared 3D model consists of 50 thousand coordinates of points.

The following formulas were applied to distort the 3D model. Along with the formulas, the values of errors and deviations that were used to distort the 3D model are also given.

1. Measurement error Δ_D of the additive constant of the distance meter during its calibration:

Figure 2.
Visualization of a prepared 3D model of a horizontal tank RHS-25.

$$D_i = \sqrt{x_i^2 + y_i^2 + z_i^2}; x_i' = x_i + \frac{x_i}{D_i}\Delta_D; y_i' = y_i + \frac{y_i}{D_i}\Delta_D; z_i' = z_i + \frac{z_i}{D_i}\Delta_D \tag{1}$$

where $\Delta_D = \pm 2$ mm.

2. Measurement error Δ_γ by scanner its rotation axis tilt if the tilt error is directed along axes x:

$$\Delta_{z_i} = x_i \cdot tg\Delta_\gamma; z_i' = z_i + \Delta_{z_i} \text{ where } \Delta_\gamma = \pm 3'. \tag{2}$$

3. Index error (of a zenith point) ν_0 at measurement of the vertical angles:

$$S_i = \sqrt{x_i^2 + y_i^2}; \Delta_{\nu_i} = D_i \cdot tg\nu_0; z_i' = z_i + \frac{S_i}{D_i}\Delta_{\nu_i}; \Delta_{S_i} = \frac{z_i}{D_i}\Delta_{\nu_i};$$
$$x_i' = x_i + \frac{x_i}{S_i}\Delta_{S_i}; y_i' = y_i + \frac{y_i}{S_i}\Delta_{S_i} \text{ where } \nu_0 = \pm 3'. \tag{3}$$

4. Deviation from the perpendicularity of the laser beam axis from the rotation axis of the mirror or tube (collimation error) C_α:

$$\Delta_{\alpha_i} = S_i \cdot tgC_\alpha; x_i' = x_i + \frac{y_i}{S_i}\Delta_{\alpha_i}; y_i' = y_i - \frac{x_i}{S_i}\Delta_{\alpha_i} \text{ where } C_\alpha = \pm 6'. \tag{4}$$

5. Deviation from the perpendicularity of the rotation axis of the mirror or tube from rotation axis of the laser scanner C_β:

$$\Delta_{\beta_i} = z_i \cdot tgC_\beta; x_i' = x_i + \frac{y_i}{S_i}\Delta_{\beta_i}; y_i' = y_i - \frac{x_i}{S_i}\Delta_{\beta_i} \text{ where } C_\beta = \pm 6'. \tag{5}$$

6. Measurement error of the absolute height of the reference point of the scanner relative to the reference point of the tank Δz_0:

$$z_i' = z_i + \Delta z_0 \text{ where } \Delta z_0 = \pm 2 \text{ mm}. \tag{6}$$

In formulas (1)–(6) the x_i, y_i, z_i are not distort coordinates, the x_i', y_i', z_i' are distort coordinates, the D_i is slope distance, the S_i is horizontal distance.

The interval capacities obtained for the distorted model were compared with the original-initial, undistorted model. The results of this comparison are shown in **Figure 3.** in a graphical form.

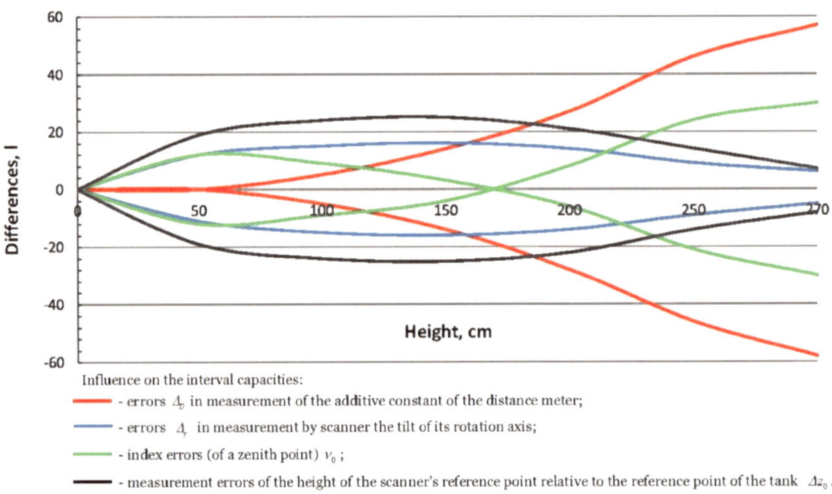

Figure 3.
Errors of the interval capacities caused by distortions of coordinates of the 3D model.

Deviations from perpendicularity in accordance with clauses 4 and 5 did not have any effect on the change in interval capacities, therefore they are absent in **Figure 3.** However, if the tank was scanned from the outside and measurements were taken from several stations, then their influence can be significant.

4.4 Tanks scanning features

Particular attention should be paid to the cleanliness of the optical surfaces of the scanner before tanks calibration and scanning. According to our research, even a small amount of dust or liquid's condensate on optical surfaces can cause a significant change in the additive constant of the distance meter.

If the tank has been hydro-tested or washed with water, then the wet walls of the tank are not an obstacle to high-quality scanning. If high humidity is inside the tank, condensation of water vapor on the optical surfaces of the scanner is not allowed. The presence of water in the bottom or in the lower part of the tank is not desirable. Water can interfere with measurements or give a large number of false points that will interfere with the preparation of a good 3D model of the tank. Measurements through the water, if they succeed, significantly distort the coordinates.

The presence of even a small amount of diesel fuel on the walls of the tank has an extremely negative effect on the results of scanning. The standard deviation of coordinate measurements (point 2, **Table 1**) can increase in 10, 20 or more times. Especially if the beam falls on the contaminated surface at an acute angle. It is very difficult or impossible to distinguish exactly where the actual surface of the tank wall is, and where is the noise from distortions by the diesel fuel slick on the tank wall.

Measurements with fuel vapors in the tank are not only hazardous to life and health. They do not provide an opportunity to make qualitative measurements. When fuel and air vapors are mixed, strong turbulence arises that is almost opaque to the laser beam. If the measurements could be carried out, then the coordinate errors can reach tens of millimeters. Such measurements must be prohibited. For these reasons, the authors of this publication are extremely skeptical about tanks scanning in accordance with [19, 20], without cleaning and degassing them. It is impossible to confirm or deny this before comprehensive comparative tests are carried out.

Definitely it is necessary to use the current additive constant to correct the coordinates on the tank wall, which is recommended to be determined independently at least once every 1–3 months.

The scanner must be calibrated at least once a year. In this case, it is recommended to determine all the characteristics and parameters of the scanner, given in **Table 1.** At least once every two – three years, it is necessary to carry out maintenance of the scanner and adjust all its parameters in the manufacturer's service center. Only the manufacturer has the opportunity to investigate in his laboratory all the characteristics and parameters of each scanner, make the necessary adjustments and settings.

5. Measurement methods of scanners during tank calibration and uncertainty sources connected with it

5.1 Features that influence the choice of the measurement method

The scanning methods for different types, sizes and purposes of tanks has a lot in common, but there are some features. Taking into account these features depend on whether the goal set in Section 2 will be achieved or not. These features include:

- location – aboveground or underground location of the tank;
- operation capability – the tank is calibrated from inside at the plant right after its production, the new one at the place of exploitations, decommissioned. The tank is calibrated from inside in case of its exploitation;
- scope – the tank is used for storage oil, oil product, liquefied gas or another liquid, for example, ammonia;
- type – a horizontal or vertical cylindrical, spherical, arbitrary geometrical.

At scanning an important condition to achieve goal set in the Section 2 is to fulfill simple but very significant rules connected with described above features.

5.2 Features of scanning through the hatch of new or decommissioned tanks

The measurement made inside the tank from a tripod is more preferable from the point of view of measurement accuracy than upside down through the hatch, as shown in **Figure 1.** On a tripod the scanner is placed almost in the center of the tank, and asymmetrically when measuring through the hatch. But to place the scanner in the center of the tank, you need to get in it, and these are additional risks to life and health. Moreover, it is extra spent time. If the characteristics and parameters of the scanner meet the recommendations given in **Table 1**, in our experience, scanning through the hatch does not lead to a significant loss of accuracy of capacity measurement. This is confirmed by the above simulations described in 4.2.

If the tank has many internal structures and equipment than many "shadows" appear on its surface (**Figure 4**). To significantly reduce their number, it is recommended to make two scans, displacing the scanner inside the hatch as much

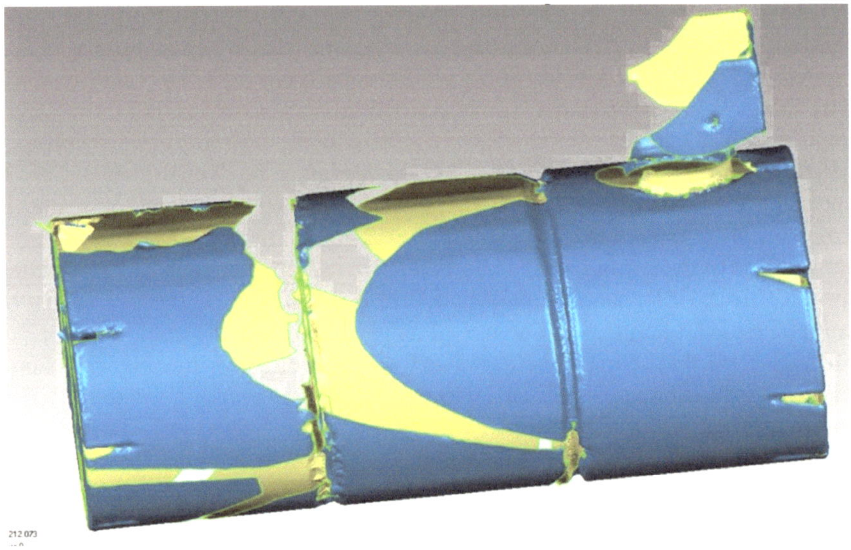

Figure 4.
Visualization of an unprepared 3D model of the horizontal tank RHS-25.

as possible. The “stitching” of two scans with the complete software for the scanner by characteristic points gives a good result.

5.3 Direct procedure for linking automatic level gauge readings to the calibration table when scanning the tank from inside

It is impossible to measure the volume and to estimate the measurement uncertainty of the liquid volume in a tank correctly, if we do not consider the pair of measuring devices “level gauge – tank” as a whole. The level gauge must be set up so that it measures the absolute height of the liquid level in the tank relative to the reference point. In other words, so that the zero of its readings coincides with the zero of the tank calibration table.

During measurements, the zero of the scanner coordinate system coincides with its reference point – the point of intersection of its horizontal and vertical axes (Section 4). During processing, it is necessary to transform the coordinates measured in the scanner coordinate system into the tank coordinate system. That is, the horizontal plane of the new coordinate system must pass through the tank reference point. It is very easy to do this with modern software if the measurements are made from the inside and the reference point is marked and clearly visible on the 3D model.

To achieve the high accuracy of measurements of the tank capacity, it is necessary that the standard uncertainty of such a link does not exceed the values given in clause 10 of **Table 1**.

Figure 5 shows a panoramic photo of the tank made by the scanner from the measuring point. Due to the fact that the scanner was installed upside down, similar to **Figure 1**, the bottom is at the top of the figure, and the roof is at the bottom. In the center of the tank, it is clearly seen the central square pipe, which supports the roof (in the photo – on the left). To the right of the pipe is a thin dark line with a dark rectangle at the end. This is a dip-tape hanging through a dip-hatch. The dip-weight of the tape touches the bottom at the dipping reference point. A horizontal plane *xy* passes through the bottom of the dip-weight, the absolute height of which is equal zero. The capacity table starts from this plane. Therefore, offset between dipping reference point (dip-point) and calibration datum point is equal zero.

Figure 5.
Recognition of the reference point of the tank on its unprepared 3D model.

After the first filling of the tank the absolute height of the level is measured by dip-tape or another material measure. This value is entered into the level gauge. Direct procedure of linking the values of the automatic level gauge to capacity table is completed. According to OIML R 85 [23], the difference between the absolute height of the level measured by the material measure and the readings of the automatic level gauge must be within ±4 mm over the entire range of level measurement for any type of tank. This value includes not only the deformation of the bottom, roof and walls of the tank, but also the uncertainty of the binding. In our opinion, this indicator for horizontal cylindrical tanks should be within ±2 mm.

5.4 Indirect procedure for linking the automatic level gauge readings to the calibration table when scanning from the inside through the tank hatch

During the repair of horizontal cylindrical tanks, remove the lid with a pipe (**Figure 1**, left), which directs the material measure to the reference point. On the inner wall of the tank, this point is almost always unmarked and it is not known exactly where it is. Without precise localization of this point, it is impossible to bind to it with the uncertainty given in item 10 of **Table 1.**

To solve this problem, it was proposed method of binding by laser level and tape. A photo of the binding procedure is shown in **Figure 1.**

The scanner is mounted on one side of the rod and the laser level (crossline) on the other. The distance from the reference point of the scanner to the end of the rod is known from metrological measurement. After tank scanning the laser beam is counted on the tape.

Further, the hatch is closed and a gage pipe is installed, which directs the material measure to the reference point. The reference height of the tank as the distance from the dipping reference point to the top of the gage pipe is measure by a material measure. A level is installed on the gage pipe (**Figure 1**, right) and the counting is carried out using a laser beam on a tape.

The absolute height of the scanner reference point in relation to the tank reference point is calculated by formula:

$$z_0 = B - d - r_1 + r_2 + c_D, \tag{7}$$

where

B is a reference height of the tank – vertical distance from the dipping reference point of the tank to the top of the gage pipe;

d is distance from the reference point of scanner to the end of the rod (**Figure 1**, left);

r_1 is reading by the laser beam of the level on a tape when it is at the end of the rod (**Figure 1**, right);

r_2 is reading by the laser beam of the level on a tape when it is at the end of the gage pipe;

c_D is the additive constant of the scanner distance meter in accordance with items 3 and 4 of **Table 1** Section 3.

The signs r_1 and r_2 in the formula are given for the case when zero of the tape is in the top (**Figure 1**). If the zero is in the bottom, it is necessary to change the signs to the opposite.

The value z_0 is added to each coordinate z_i. It is easy to complete by software bundled with the scanner during preliminary processing of the measurement results.

Before start-up a horizontal cylindrical tank with a capacity of up to 50 m^3 after its calibration, it is recommended to check the correctness of measuring the absolute height of the scanner's reference point relative to the tank's reference point. To do this, it is necessary to pour a known volume of fuel into an empty tank through a standard batch meter or fuel dispenser in accordance with ISO 4269 [15], for example, for a horizontal tank, 100 or 200 liters, and measure the level of fuel in the tank using a material measure. The measured level must correspond to one in the capacity table within limits pointed out in the item 10 of **Table 1**. In justified cases, it is recommended to shift the table so that the measured level in it coincides with the poured controlled volume of the liquid.

5.5 An indirect procedure for linking the automatic level gauge readings to the capacity table when scanning the tank from the outside

An indirect procedure is used to calibrate tanks for liquefied gas that are operated under pressure and for some other tanks that are operated at atmospheric pressure. Link to the reference point is fulfilled by the tank edge where the automatic level gauge is mounted. As a rule, for such tanks measurements by the material measure are not provided even if the tank is not out of service. The zero of the level gauges reading during calibration must align with the stock on the level gauge that connects it to the stock on the tank, taking into account the thickness of the gasket. Before scanning, the stock must be marked with the brand recommended by the scanner manufacturer.

In GOST 8.655 [11, 17] for spherical tanks it was proposed to take the lowest point of the approximating sphere as the reference point of the tank. The absolute height of the stock above the reference point is the sum of the radius of the tank and the distance from the stock to the center of the tank, which are calculated from the results of the approximation (Section 7). This makes it possible to indirectly link the capacity table to the readings of the level gauge with the uncertainty given in the clause 10 of **Table 1.**

6. Preparation of a 3D model of a tank to calculate its capacity: unprepared and prepared 3D model

In general, the operations of preparation to calculate the capacity of different types of tanks coincide, but may differ somewhat depending on the features given in subsection 5.1.

If the tank was scanned from one point (station), then this is an unprepared 3D model of the tank (**Figure 4**). If the tank was scanned from several points (stations), then several scans are combined into a single unprepared 3D model of the tank by the software bundled with the scanner. It is impossible to use such a model to calculate interval capacities. It must be properly prepared.

A prepared 3D model of the tank (**Figure 2**) is a file of spatial coordinates of points on its surface available for visualization and editing with special software, as well as other calculations, for example, approximation, triangulation, calculation of capacity.

To create a prepared for calculation 3D model the following procedures should be carried out:

1. Separate scans should be combined into one unprepared 3D model (**Figure 4**). The empty spaces that are closed for measurement by internal constructions of the tank are visible on such model. The constructions that do not belong to the tank walls are also visible;

2. The point coordinates should be corrected by amendment that takes in account the additive constant of the scanner distance meter;

3. During internal measurements the 3D model should be turned over if scanning was performed "head down" by the scanner (**Figure 1**). During external measurement this option is absent;

4. Align the horizontal plane xy with the reference point of the tank (subsection 5.4) and check the alignment of the axis z with the plumb line;

5. Change over the axes x and y, if the program will be used that have another axes orientation;

6. To measure geometrical sizes and the absolute height z, in a new coordinate system, internal constructions and equipment. During external measurements it is possible to use materials of the previous calibration or drafts of the tank;

7. Delete from a 3D model all points that do not belong to tank walls including internal constructions and equipment;

8. Using the special function of the software reduce the number of points on the tank wall so that remained from 40 to 100 thousand, but they should evenly cover the walls;

9. To open 3D model so as necessary for calculation a tank capacity, for example, at approximation of the horizontal tanks the axis x is oriented in a horizontal plane LSM approximately parallel to the tank axes;

10. If it is supposed to use triangulation method for capacity calculation (Section 7), then special software generates points where they are absent (**Figure 2**). For LSM approximation method (Section 7) it is useless.

The internal constructions and equipment are presented as simple geometrical shapes – parallelepiped or cylindrical. For each shape directly on the 3D model it is measured:

- geometrical sizes – length, width and height or length and diameter;
- axis tilt of the parallelepiped or cylinder to the plumb line;
- the absolute height of the lowest and the highest point;
- for the vertical tanks it is measured the absolute height of the courses;
- for the vertical tanks by the floating roof or pontoon it is measured their geometrical sizes and the absolute height of their lowest part.

These data are sufficient to compose a capacity table for each such geometric shape from the minimum to the maximum absolute height, as described in [11–13]. These local capacity tables of constructions and equipment are added as corrections to the main tank capacity table.

For tanks with a floating roof or pontoon the correction is calculated from the absolute height at the moment the liquid touches them to the absolute height of their emersion.

7. Estimation of the interval capacities of the tanks and their uncertainty

7.1 Evaluation of the interval capacities by the results of approximation and triangulation

A tank should be considered as a single spatial surface where a spatial coordinates of each point of the prepared 3D model are known with a required uncertainty in a single spatial coordinate system. Methods of calculation of the interval capacity can be considered as strict if they mathematically correct use all the points of the prepared 3D model. In the standards [2–10] it is applied non-strict models where separate sections are used to measure tank radius. As a result, some of the valuable information about the shape and size of the tank is not used. This leads to a latent (it is almost impossible to estimate) increase in the calculation in the interval capacities uncertainty.

In our opinion, at the moment, two strict methods have been developed to calculate the interval capacities of tanks by the coordinates of points obtained by the laser scanning method. One is based on the approximation of a spatial surface of a regular geometric shape, and the other is on the triangulation of the surface. Each of them consists of two stages.

The approximation of the surface by the least square method. At the first stage, the geometrical parameters of the tank are calculated from the coordinates of all points on the surface – the mean radius of the spherical tank GOST 8.655 [11, 17] or the mean radius and the axis tilt of the vertical cylindrical tank GOST 8.656 [12, 16] the mean radius, the axis tilt and the length of the horizontal cylindrical tank GOST 8.659 [13]. The radial deviations of the real surface of the tank relative to the approximating one are also calculated.

At the second stage, the volume of the segments of such an approximating surface is calculated below the specified horizontal planes. This capacity comprises correction for relief based on the calculated radial deviations. The evaluation of the uncertainty of interval capacities should be based on the fundamental documents JCGM 100 [24] and JCGM 102 [25]. The process of evaluation the interval capacities and their uncertainties by this method is described in [11–13, 16, 17] and improved in subsection 7.2.

The triangulation of the tank surface by the Delone's method [26] is fulfilled at the first stage. We did not consider it is necessary to refer to a big number of sources on this topic except the first publication of the author. Delone's method is not the topic to discuss in this publication.

At the second stage, the area of a closed polygon is calculated, which lies at the intersection of the horizontal plane and the plane of the triangles that it crossed. At the intersection of the sides of each triangle and the horizontal plane, the coordinates of the points are calculated. As a result, these coordinates are used to calculate the area of a closed polygon. The complexity of this algorithm is that the program must subtract from the area of this polygon (conditionally "sea") the area of all "islands". But add the area of all the "lakes" on the "islands". Then, subtract the areas of all the "islands" on the "lakes", etc.

The interval capacity is calculated from the areas of two polygons and the increment in height between the areas.

Both of these two strict methods have their own advantages and disadvantages. So, for the approximation method the uniform coverage of the entire surface by points is useless, simply they must be quite enough. Part of the surface may not be covered by points at all. Even a small number of points (for example 3 000) evenly spaced on the cylindrical surface of the vertical tank give an acceptable capacity uncertainty. However, this method is not applicable to arbitrary geometrical shape tanks, for example ships' tanks.

The triangulation method allows to calculate tank capacity of any arbitrary shape, however, point on the tank surface should be many (not less than 30–50 thousand). They must be placed evenly otherwise the program will not connect them in triangles.

It is impossible to calculate the interval capacities of the tank reliably if part of its surface is not covered by triangles (**Figure 4**). It is necessary to generate on these sections the additional points by the special program (**Figure 2**). This increases the uncertainty of a capacity calculation. The triangles should be not elongated and/or large as they will not follow the real shape of the surface. It is good if the program adds on the points on the bending lines, for example, on the line, where the flat bottom is linked up the cylindrical part. If this is not done, then the real sharp bend line will be smoothed out. It negatively affects on the uncertainty of the interval capacities.

The advantages and disadvantages of these methods are not limited to those mentioned above, however, they have clear advantages over non-strict methods. Theoretically, the capacity of the same tank for which the same 3D model is used, processed by different programs based on strict methods, should coincide. However, in practice this is not the case.

Preliminary results of comparisons of tanks capacity calculated by different software are described in [14].

7.2 Evaluation of the geometrical parameters and capacity of spherical and cylindrical surfaces and their uncertainties

7.2.1 Identifying the problem

The problem is that at approximation it is necessary to determine spatial position and orientation of the tank surface toward horizontal plane xy, mentioned above, otherwise, it is impossible to calculate the interval capacities correctly. The spatial position of the spherical surface identifies by 3D coordinates of its center, and the cylindrical surface by the coordinates of the point on its axis. The spatial orientation of the cylindrical surface of the vertical tank is identified by the tilt of its axis to the plumb line in the coordinate planes xz and yz. The spatial orientation of the cylindrical surface of the horizontal tank is identified by the tilt of its axis to horizontal plane xy and rotation to the plane xz.

7.2.2 Evaluation of a spherical and cylindrical surface geometrical parameters

The measurement models binding points coordinates on the surfaces and their geometrical parameters are given by Eq. (8):

$$\phi_i(x_i, y_i, z_i, \tau_1 \dots \tau_k) = 0 \tag{8}$$

For spherical surface the measurement model (8) is:

$$R = \sqrt{(x_i - x_o)^2 + (y_i - y_o)^2 + (z_i - z_o)^2}, \tag{9}$$

For vertical and horizontal cylindrical surface the measurement model (8) are:

$$R = \sqrt{\frac{(x_i - x_o - z_i \cdot \eta_x)^2}{1 + \eta_x^2} + \frac{\left(y_i - y_o - z_i \cdot \eta_y\right)^2}{1 + \eta_y^2}}; \tag{10}$$

$$R = \sqrt{\frac{(z_i - z_o - x_i \cdot \eta_z)^2}{1 + \eta_z^2} + \frac{\left(y_i - y_o - x_i \cdot \eta_y\right)^2}{1 + \eta_y^2}} \tag{11}$$

where:

x_i, y_i, z_i are the horizontal coordinates and the absolute height of the points on the surface (where $i = 1 \dots n$), which are measured by the 3-D instruments;

$\tau_1 \dots \tau_k$ are the defined geometrical parameters;

R is the mid radius of the spherical or cylindrical surface;

x_o, y_o, z_o are the coordinates of center for the spherical surface;

x_o, y_o are the horizontal coordinates of the point on the surface's axis if $z_o = 0$ for the vertical cylindrical surface;

z_o, y_o are the absolute height the and horizontal coordinate of the point on the surface's axis if $x_o = 0$ for the horizontal cylindrical surface;

$$\eta_x = tg\beta_x; \eta_y = tg\beta_y; \eta_z = tg\beta_z \tag{12}$$

where

$\beta_x, \beta_y, \beta_z$ are the tilt of the cylindrical surface axis – the angles in the plane projection coordinates xz, yz and xy;

k is the quantity of the determined geometrical parameters of the spherical or cylindrical surfaces;

n is the quantity of points at the spherical or cylindrical surfaces at which the coordinates are determined.

For surfaces the absolute height is the vertical distance from any horizontal flatness to the point with number i.

Due to surface roughness and coordinate measurement uncertainties, Eq. (8), (9), (10) and (11) is not fulfilled. That is why, to evaluate the determined geometrical parameters of the spherical, cylindrical and other surfaces, it is necessary to establish equations of corrections, which we obtain by the partial derivation of the measurement model (8) by the measured coordinates and defined parameters:

$$\frac{\partial\phi_i}{\partial x_i}v_{x_i}+\frac{\partial\phi_i}{\partial y_i}v_{y_i}+\frac{\partial\phi_i}{\partial z_i}v_{z_i}=\frac{\partial\phi_i}{\partial\tau_1}\delta\tau_1+\dots+\frac{\partial\phi_i}{\partial\tau_j}\delta\tau_j+\dots+\frac{\partial\phi_i}{\partial\tau_k}\delta\tau_k+l_i \text{ or } \vartheta_i$$
$$=\frac{\partial\phi_i}{\partial\tau_1}\delta\tau_1+\dots+\frac{\partial\phi_i}{\partial\tau_j}\delta\tau_j+\dots+\frac{\partial\phi_i}{\partial\tau_k}\delta\tau_k+l_i \tag{13}$$

where:

$v_{x_i}, v_{y_i}, v_{z_i}$ are corrections to the measured coordinates of the point with number i on the surface;

ϑ_i is the radial deviation of the real surface perpendicular to the approximating surface;

$\delta\tau_1$, $\delta\tau_j$, $\delta\tau_k$ are corrections to the initial values of the defined parameters of the surface $\tau_1^0 \dots \tau_k^0$;

$l_i=\phi_i\left(x_i, y_i, z_i, \tau_1^0 \dots \tau_k^0\right)$ is the constant term of the equation of corrections.

In the matrix, the parametric equation of the corrections system (13) gives:

$$A\cdot V=B\cdot\delta\tau+l \text{ or } \vartheta=B\cdot\delta\tau+l \tag{14}$$

where:

A is the matrix of the partial derivatives from the measurement model (8) by the measured coordinate points;

V is the correction matrix to the measured coordinate points;

ϑ is the diagonal matrix of the radial deviation of the real surface from the approximation;

B is the partial derivative matrix from the measurement model (8) by the determined geometrical parameters;

$\delta\tau$ is the correction vector to the initial values of the determined parameters;

l is the constant terms vector of the correction equation.

Then, Eq. (13) for spherical surfaces are given by:

$$\vartheta_i^{(g)}=b_{i1}\cdot\delta R^{(g)}+b_{i2}\cdot\delta x_o^{(g)}+b_{i3}\cdot\delta y_o^{(g)}+b_{i4}\cdot\delta z_o^{(g)}+l_i^{(g)} \tag{15}$$

where:

$$b_{i1}=1; a_{i1}=b_{i2}=-\frac{x_j-x_o^{(g)}}{R^{(g)}}; a_{i2}=b_{i3}=-\frac{y_i-y_o^{(g)}}{R^{(g)}}; a_{i3}=b_{i4}=-\frac{z_i-z_o^{(g)}}{R^{(g)}}; \tag{16}$$

$$l_i^{(g)}=R^{(g)}-\sqrt{\left(x_i-x_o^{(g)}\right)^2+\left(y_i-y_o^{(g)}\right)^2+\left(z_i-z_o^{(g)}\right)^2}. \tag{17}$$

where:

$R^{(g)}$, $x_o^{(g)}$, $y_o^{(g)}, z_o^{(g)}$ are the initial values of the defined geometrical parameters of the spherical surface;

g is the number of approximation.

Eq. (13) for vertical cylindrical surfaces are given by:

$$\vartheta_i^{(g)}=b_{i1}\cdot\delta R^{(g)}+b_{i2}\cdot\delta x_o^{(g)}+b_{i3}\cdot\delta y_o^{(g)}+b_{i4}\cdot\eta_x{}^{(g)}+b_{i5}\cdot\eta_y{}^{(g)}+l_i^{(g)} \tag{18}$$

$$b_{i1}=1; b_{2i}=\frac{x_i-x_o-\eta_x\cdot z_i}{R}; b_{3i}=\frac{y_i-y_o-\eta_y\cdot z_i}{R};$$
$$b_{4i}=\frac{x_i-x_o-\eta_x\cdot z_i}{R}z_i; b_{5i}=\frac{y_i-y_o-\eta_y\cdot z_i}{R}z_i; \tag{19}$$

$$l_i^{(g)} = R^{(g)} - \sqrt{\left(x_i - x_o^{(g)} - \eta_x^{(g)} \cdot z_i\right)^2 + \left(y_i - y_o^{(g)} - \eta_y^{(g)} \cdot z_i\right)^2} \tag{20}$$

The dual weight of the radial deviation ϑ_i and the radial deviation weight are given by:

$$\frac{1}{w_i} = Q_i = |a_{i1}a_{i2}a_{i3}| \cdot \begin{vmatrix} q_{x_i} K_{x_i y_i} K_{x_i z_i} \\ K_{y_i x_i} q_{y_i} K_{y_i z_i} \\ K_{z_i x_i} K_{z_i y_i} q_{z_i} \end{vmatrix} \cdot \begin{vmatrix} a_{i1} \\ a_{i2} \\ a_{i3} \end{vmatrix}^T \tag{21}$$

$$w_i = \frac{1}{Q_i} = \left(A_i q_{xyz}^i A\right)^{-1} \tag{22}$$

where:

$q_{x_i}, q_{y_i}, q_{z_i}$ are the dual weights of the defined coordinates;

$K_{x_i y_i}, K_{x_i z_i}, K_{y_i x_i}, K_{y_i z_i}, K_{z_i x_i}, K_{z_i y_i}$ are the covariance (correlation) moments;

q_{xyz}^i is the covariance (correlation) coordinate matrix;

$A_i = |a_{i1}a_{i2}a_{i3}|$ is the partial derivative matrix of Eq. (8) by the measured coordinates (see Eqs. (13) and (14)).

Taking into account that the parametric equations of corrections (13) (for spherical surfaces (15), for vertical cylindrical surfaces (18)) are much greater than the determined geometrical parameters, one may build a normal equation system, which in the matrix, taking into account that $B^T \cdot W \cdot \vartheta = 0$, gives:

$$B^T \cdot W \cdot B \cdot \delta\tau + B^T \cdot W \cdot l = 0 \text{ or } N \cdot \delta\tau + L = 0 \tag{23}$$

Corrections to the initial values of the geometrical parameters determined in the matrix are obtained by solving the system of linear Eq. (23) by the formula:

$$\delta\tau = -N^{-1} \cdot L = -Q \cdot L \tag{24}$$

where:

$N^{-1} = Q$ is the inverse matrix to the normal equation matrix.

The evaluation by the least square method using the covariance matrix point coordinates at the surface is fulfilled under the following conditions:

$$\vartheta \cdot W \cdot \vartheta^T = min \text{ or } \sum_{i=1}^{n} w_i \cdot \vartheta_i^2 = min \tag{25}$$

where w_i are the weights of the radial deviations.

The standard radial deviation σ_ϑ of the real surface from the approximation is calculated by the formula:

$$\sigma_\vartheta = \sqrt{\frac{\vartheta \cdot W \cdot \vartheta^T}{n-k}} = \sqrt{\frac{\sum_{i=1}^{n} W \cdot \vartheta_i^2}{n-k}} \tag{26}$$

The adjusted (defined) parameters, their covariance matrix and the standard uncertainties (standard deviation) of the main defined surface geometrical parameter – mid internal radius and center coordinates – are:

$$R^{(g+1)} = R^{(g)} + \delta R^{(g)}; x_o^{(g+1)} = x_o^{(g)} + \delta x_o^{(g)}; y_o^{(g+1)} = y_o^{(g)} + \delta y_o^{(g)}; z_o^{(g+1)} = z_o^{(g)} + \delta z_o^{(g)}; \tag{27}$$

$$K_{\delta\tau} = \sigma_\vartheta^2 \cdot Q; \tag{28}$$

$$u_A(R) = \sigma_R = \sigma_\vartheta \sqrt{Q_{11}}; u_A(x_o) = \sigma_{x_o} = \sigma_\vartheta \sqrt{Q_{22}};$$
$$u_A(y_o) = \sigma_{y_o} = \sigma_\vartheta \sqrt{Q_{33}}; u_A(z_o) = \sigma_{z_o} = \sigma_\vartheta \sqrt{Q_{44}} \tag{29}$$

where:

Q_{11}, Q_{22}, Q_{33} and Q_{44} are the diagonal components of the inverse matrix to the normal equation matrix.

The standard radial deviation σ_ϑ, calculated by formula (26), comprises not only measurement uncertainties. It also comprises real surface deviations from the mathematically correct shape. They appear when the surface is manufactured and due to deformations caused by its use. Based on their experience in calibration, the author wishes to state that the second component exceeds the first for tanks surface by 5–10 times.

7.2.3 Measurement model for the spherical and cylindrical surface total and interval capacities

The adjusted spherical, cylindrical or other surfaces' interval capacity below absolute height H_f is:

$$\overline{V}_{H_f} = V_{H_f} + \Delta V_{H_f} \tag{30}$$

where

V_{H_f} is the interval capacity for surface that was calculated by adjusted parameters of surface;

ΔV_{H_f} is the terrain corrections to capacity below height H_f.

The measurement model of the total and interval capacities of the spherical surface is:

$$V_{H_f} = \psi_V(\tau_1 \dots \tau_k, H_f) = \pi \cdot H_f^2 \cdot (R - H_f/3), \tag{31}$$

and of the vertical cylindrical surface is:

$$V_{H_f} = \pi \cdot R^2 \cdot \sqrt{1 + \eta_x^2 + \eta_y^2} \cdot H_f = \pi \cdot R^2 \cdot k_\eta \cdot H_f \tag{32}$$

where:

H_f is the absolute height of the horizontal plane (for example, liquid) relative the bottom point of the approximation spherical surface or reference plane for cylindrical surface;

R is the adjusted radius of the spherical or cylindrical surface;

f is current number for which the interval capacity is calculated;

π= 3.1415926 ....

Taking into account a large number of points whose coordinates are determined by the method of laser scanning on the surface, it is proposed to calculate the terrain corrections ΔV_{H_f} using formulas (33) and (34):

$$\overline{\vartheta}_{H_f} = \frac{\sum_{i=1}^{n_{H_f}} \vartheta_i}{n_{H_f}}; \Delta V_{H_f} = S_{H_f} \cdot \overline{\vartheta}_{H_f}; \tag{33}$$

where for the spherical surface

$$S_{H_f} = \psi_S(\tau_1 \dots \tau_k, H_f) = 2 \cdot \pi \cdot \overline{R} \cdot H_f \tag{34}$$

for the vertical cylindrical surface

$$S_{H_f} = \pi \cdot R \cdot \left(1 + \sqrt{1 + \eta_x^2 + \eta_y^2}\right) \cdot H_f = \pi \cdot R \cdot (1 + k_\eta) \cdot H_f \tag{35}$$

where:
S_{H_f} is the surface's area below height H_f;
$\overline{\vartheta}_{H_f}$ is the mid radial deviation of the surface's wall below height H_f (ϑ_i is calculated by (15) or (18));
n_{H_f} is the number of points on the surface below height H_f, the coordinates of which were determined.

7.2.4 Evaluation of the total and interval capacities uncertainty of the spherical and cylindrical surfaces

The evaluation of the standard deviation (A-type standard uncertainty) unimproved by the spherical surface interval capacity below height H_f is calculated considering correlation by formula (36):

$$u_A^2\left(V_{H_f}\right) = \sigma_{V_{H_f}}^2 = F_{V_{H_f}} \cdot K_{\delta\tau} \cdot F_{V_{H_f}}^T = \sigma_\vartheta^2 \cdot F_{V_{H_f}} \cdot Q \cdot F_{V_{H_f}}^T \tag{36}$$

where for the spherical surface:

$$F_{V_{H_f}} = \left| \frac{\partial \psi_V}{\partial R} \; \frac{\partial \psi_V}{\partial x_o} \; \frac{\partial \psi_V}{\partial y_o} \; \frac{\partial \psi_V}{\partial z_o} \right| = \left| \pi H_f^2 \;\; 0 \;\; 0 \;\; \pi H_f (2R - H_f) \right| \tag{37}$$

is the partial derivatives vector from function (31) by the spherical surface's geometrical parameters R and z_o.
For the cylindrical surface:

$$F_{V_{H_f}} = \left| \frac{\partial \psi_V}{\partial R} \; \frac{\partial \psi_V}{\partial x_o} \; \frac{\partial \psi_V}{\partial y_o} \; \frac{\partial \psi_V}{\partial \eta_x} \; \frac{\partial \psi_V}{\partial \eta_y} \right| = \left| 2\pi R k_\eta H_f \quad 0 \quad 0 \quad \pi R^2 \frac{\eta_x}{k_\eta} H_f \quad \pi R^2 \frac{\eta_y}{k_\eta} H_f \right| \tag{38}$$

is the partial derivatives vector from function (32) by the cylindrical surface's geometrical parameters R, η_x and η_y.
Formula (36) excludes correlation for spherical surface gives:

$$u_A\left(V_{H_f}\right) = \pi \sqrt{H_f^4 u_A^2(R) + H_f^2 (2R - H_f)^2 u_A^2(z_o)} \tag{39}$$

Formula (36) excludes correlation for vertical cylindrical surface gives:

$$u_A\left(V_{H_f}\right) = 2\pi R H_f \sqrt{k_\eta^2 u_A^2(R) + \left(\frac{R}{2k_\eta}\right)^2 \left(\eta_x^2 u_A^2(\eta_x) + \eta_y^2 u_A^2\left(\eta_y\right)\right)} \tag{40}$$

The evaluation of the standard deviation (A-type standard uncertainty) of the spherical and vertical cylindrical surfaces' area below height H_f is calculated considering correlation by formula (41):

$$u_A^2\left(S_{H_f}\right) = \sigma_{S_{H_f}}^2 = F_{S_{H_f}} \cdot K_{\delta\tau} \cdot F_{S_{H_f}}^T = \sigma_\vartheta^2 \cdot F_{S_{H_f}} \cdot Q \cdot F_{S_{H_f}}^T \tag{41}$$

where for the spherical surface:

$$F_{S_{H_f}} = \left| \frac{\partial \psi_S}{\partial R} \; \frac{\partial \psi_S}{\partial x_o} \; \frac{\partial \psi_S}{\partial y_o} \; \frac{\partial \psi_S}{\partial z_o} \right| = \left| 2\pi H_f \quad 0 \quad 0 \quad 2\pi R \right| \tag{42}$$

is the partial derivatives vector from function (31) by the spherical surface's geometrical parameters R and z_o.

For the cylindrical surface:

$$F_{S_{H_f}} = \left| \frac{\partial \psi_S}{\partial R} \; \frac{\partial \psi_S}{\partial x_o} \; \frac{\partial \psi_S}{\partial y_o} \; \frac{\partial \psi_S}{\partial \eta_x} \; \frac{\partial \psi_S}{\partial \eta_y} \right| = \left| \pi(1+k_\eta)H_f \quad 0 \quad 0 \quad \pi R\frac{\eta_x}{k_\eta}H_f \quad \pi R\frac{\eta_y}{k_\eta}H_f \right| \tag{43}$$

is the partial derivatives vector from function (32) by the vertical cylindrical surface's geometrical parameters R, η_x and η_y.

Formula (41) excluding correlation for spherical surface gives:

$$u_A\left(S_{H_f}\right) = \pi\sqrt{(2H_f)^2 u_A^2(R) + (2R)^2 u_A^2(z_o)} \tag{44}$$

Formula (41) excluding correlation for cylindrical surface gives:

$$u_A\left(S_{H_f}\right) = \pi H_f \sqrt{(1+k_\eta)^2 u_A^2(R) + \left(\frac{R}{k_\eta}\right)^2 \left(\eta_x^2 u_A^2(\eta_x) + \eta_y^2 u_A^2\left(\eta_y\right)\right)} \tag{45}$$

The standard deviation of the mid radial deviation of the surface's wall below height H_f, which is given in first formula (33), is calculated by formula (46):

$$u_A\left(\overline{\vartheta}_{H_f}\right) = \sigma_{\overline{\vartheta}_{H_f}} = \sqrt{\frac{\sum_{i=1}^{n_{H_f}} \left(\vartheta_i - \overline{\vartheta}_{H_f}\right)^2}{n_{H_f} \cdot \left(n_{H_f} - 1\right)}} \tag{46}$$

The standard deviation (A-type uncertainty) of the terrain correction ΔV_{H_f} is calculated for all type of surfaces by formula (47):

$$u_A\left(\Delta V_{H_f}\right) = \sqrt{\overline{\vartheta}_{H_f}^2 \cdot u_A^2\left(S_{H_f}\right) + S_{H_f}^2 \cdot u_A^2\left(\overline{\vartheta}_{H_f}\right)} \tag{47}$$

The A-type uncertainty of the adjusted surface interval capacity below height H_f is calculated using formula (48):

$$u_A\left(\overline{V}_{H_f}\right) = \sqrt{u_A^2\left(V_{H_f}\right) + u_A^2\left(\Delta V_{H_f}\right)} \quad (48)$$

Thus, it has been developed a method for evaluation the geometrical parameters and interval capacities of spherical, vertical and horizontal cylindrical tanks, as well as evaluation of their uncertainty. It provides adequate comparable results regardless of the size and shape of the surface, deformation of its surface, measurement accuracy, as well as the location and number of points which coordinates were determined on the surface.

7.3 Corrections to the tank capacity: uncorrected and corrected capacity

For accurate measurement of the liquid level during commercial and tax operations, the internal accounting and inventory it is very important to insert corrections to the tank capacity properly. These are small values, but they are systematic and can significantly contribute to the uncertainty of a liquid volume measurement.

Corrections are inserted for all types of tanks:

- for bringing the interval capacity of the tank to a temperature 15°C;
- for presence the internal constructions and equipment.

If the tank was calibrated from outside then corrections are inserted for tank wall and paint thickness.

For tanks that are operated under an atmospheric pressure, corrections are inserted for:

- hydrostatic pressure of the liquid in the tank during calibration;
- hydrostatic pressure of the liquid on walls of the tank at its operation;
- For sealed tanks that are operated under excess pressure, instead of corrections for hydrostatic pressure it is inserted the corrections for:
- the excess pressure that was in the tank at its calibration;
- the excess pressure that will be in the tank at its operation.

For example, liquefied gas tanks are operated under pressure. Therefore, the radius of the tank must be corrected for the pressure inside the tank at calibration and the average pressure in the tank at its operation.

Tanks are used not only for commercial operations but also for internal account and inventory. For such operations, it is important to know as accurately as possible the entire volume of liquid in the tank. Therefore, for vertical cylindrical tanks it is inserted corrections for:

- volume of liquid replaced by the floating roof or pontoon;
- "deadwood" volume in the lower part of the tank.

For each tank, the absolute height of the liquid level must be established, below which the commercial operations are not recommended, since the uncertainty of the capacity there is greater than the standardized one. For example, the capacity of a vertical cylindrical tank below a drain pipe or below the height of a floating roof or pontoon.

Mentioned above corrections should not be inserted manually. Their introduction should be carried out by carefully tested software for calculation the interval capacities.

8. Comparison of the interval capacities of the tanks calculated by the laboratories' own software

The best way to prepare a laboratory for an accreditation according to ISO/IEC 17025 [27] is to test qualification of the laboratory in accordance with ISO/IEC 17043 [28] and EA-2/14 [29]. Processing of the test results to fulfill taking into account the requirements of ISO 13528 [30]. This will help to solve the problem of identifying sources of errors made by laboratories.

Taking into account many sources of generation the internal capacities' uncertainty and difficulties in organization such kind of work it is proposed to fulfill it in four stages – from simple to complex:

a. The participants calculate the interval capacities of the prepared by the provider 3D model of the tank without inserting any corrections;

b. The participants calculate the interval capacities of the prepared by the provider 3D model of the tank with inserting corrections in accordance with the provider's protocol;

c. The participants prepare 3D model of the tank that was scanned by the provider with the subsequent calculation of the interval capacities according to (a) and/or (b);

d. Scanning of the tank prepared by the provider with its own scanner with the subsequent calculation of the interval capacities according to (a), (b) and (c).

Organization of the first three stages will not be too difficult and expensive to the provider and the laboratories. Only after formation of group of laboratories that have successfully pass the first three stages, it will be reasonably to carry out the fourth stage that is relatively expansive.

9. Conclusions

1. The laser scanning and special software, in the practice of calibration of the different types of the tanks, allow to achieve the main goal – to reduce uncertainty in definition of the tanks' interval capacities:

    - vertical cylindrical – to 0,05–0,1% (OIML R71 [1] requirement – 0,2% and 0,3% for tilted tanks);

    - horizontal cylindrical – to 0,1–0,2% (OIML R71 [1] requirement – 0,3%);

    - spherical – to 0,07–0,15% (OIML R71 [1] requirement – 0,5%);

- irregular geometrical shape – to 0,1–0,3% (OIML R71 [1] requirement – 0,5%).

2. Significantly shortening of measurement time using laser scanning of tanks:

   - vertical and horizontal cylindrical and spherical is up to 10–40 minutes for internal and external measurements correspondingly;

   - irregular geometrical shape is up to 30–120 minutes for internal measurements.

3. Significantly shortening the time of processing and creating of calibrating table of tank based on laser scanning data using special software:

   - vertical and horizontal cylindrical and spherical is up to 20–60 minutes for internal and external measurements correspondingly;

   - irregular geometrical shape is up to 60–180 minutes for the internal measurements.

4. To achieve the target goal the calibration laboratory must:

   - have a device that conform to the target goals and keep it in a good technical condition;

   - strictly follow the measurement procedure during the tanks' scanning;

   - have special software based on the rigorous mathematical methods. It must pass the comprehensive tests against the target goal;

   - strictly obey the procedure of processing the scanning results and calculation of the interval capacities and evaluation of their uncertainty;

   - take part in the interlaboratory comparisons of the results of the capacity calculation of the same 3D model of the tanks.

5. The proposed comparisons are needed to the laboratories in order to identify not only errors included in the mathematical model of calculations and/or made at programming, but also to test the ability to prepare a 3D model for calculations correctly and to calculate the capacity of the existing program correctly. Participation of the laboratories in comparisons can be considered as a process of validation of calculation methods and software.

6. There is a need to create new international standards that describe the entire calibration process for all types of tanks based on the principles described in this publication

Author details

Oleksandr Samoilenko* and Volodymyr Zaets
State Enterprise "Ukrmetrteststandard", Kyiv, Ukraine

*Address all correspondence to: asam@ukrcsm.kiev.ua

References

[1] OIML R 71:2008 Fixed storage tanks. General requirements. Paris – France: Bureau International de Métrologie Légale; 2008. Available from: https://www.oiml.org. p. 22

[2] ISO 7507-1:2003 Petroleum and liquid petroleum products – Calibration of vertical cylindrical tanks – Part 1: Strapping method. Switzerland: ISO; 2003. p. 62

[3] ISO 7507-4:2010 Petroleum and liquid petroleum products – Calibration of vertical cylindrical tanks – Part 4: Internal electro-optical distance-ranging method. Switzerland: ISO; 2010. p. 20

[4] ISO 7507-5:2000 Petroleum and liquid petroleum products – Calibration of vertical cylindrical tanks – Part 5: External electro-optical distance-ranging method. Switzerland: ISO; 2000. p. 11

[5] ISO 12917-2:2002 Petroleum and liquid petroleum products – Calibration of horizontal cylindrical tanks – Part 2: Internal electro-optical distance-ranging method. Switzerland: ISO; 2002. p. 14

[6] EN ISO 8311:2013 Refrigerated hydrocarbon and non-petroleum based liquefied gaseous fuels – Calibration of membrane tanks and independent prismatic tanks in ships – manual and internal electro-optical distance-ranging methods. Switzerland: ISO; 2013. p. 39

[7] GOST P 8.994 — 2020 Steel horizontal cylindrical tanks. Calibration procedure using the electron optical method. Moscow: Standardinform; 2021. p. 69

[8] GOST P 8.996 — 2020 Steel upright cylindrical tanks. Calibration procedure using the electron optical method. Moscow: Standardinform; 2021. p. 89

[9] API Standard. Manual of Petroleum Measurement Standards. Chapter 2 – Tank Calibration. Section 2D – Calibration of Upright Cylindrical Tanks Using the Internal Electro-optical Distance Ranging Method. USA Standard. American Petroleum Institute; 2003. p. 11.

[10] API Standard 2552 (R2018). Method for measurement and calibration of spheres and spheroids. USA Standard. American Petroleum Institute; 1965. p. 17.

[11] GOST 8.655-2016 The reservoirs for fluid gas steel spherical. Methods of the verification (calibration) by geometrical method with the use of the geodetic instruments (DSTU 7474 IDT). Minsk: Euro-Asian council for standardization, metrology and certification; 2016. p. 75

[12] GOST 8.656-2016 The vertical measurement reservoirs. Methods of the verification (calibration) by geometrical method with the use of the geodetic instruments (DSTU 7473 IDT). Minsk: Euro-Asian council for standardization, metrology and certification; 2016. p. 112

[13] GOST 8.659-2016 The reservoirs steel cylindrical horizontal. Methods of the verification (calibration) by geometrical method with the use of the geodetic instruments (DSTU 7475 IDT). Minsk: Euro-Asian council for standardization, metrology and certification; 2016. p. 108

[14] O. Samoylenko, V. Zaets Interlaboratory comparisons of tank capacity calculated by laboratories' own software – key stage of their proficiency testing OIML bulletin, – Paris.: Volume LIX, Number 4, October 2018. pp. 5-11

[15] ISO 4269:2001 Petroleum and liquid petroleum products – Tank calibration by liquid measurement – Incremetal method using volumetric method. Switzerland: ISO; 2001. p. 26

[16] O. Samoylenko, V. Zaets Evaluation by least square method of geometrical parameters and capacity of all tank tapes by the results of laser scanning – Paris.: OIML Bulletin, Volume LVI, Number 3, July 2015. pp. 14-21

[17] O. Samoylenko, V. Zaets Evaluation of the uncertainties of the geometrical parameters and capacity of spherical tanks – Paris.: OIML Bulletin, Volume LVII, Number 4, October 2016. pp. 10-16

[18] Wang Jintao, Tong Lin, Ma Xinyu and Zheng He The Non-contact Precision Measurement for Spherical Metrology Tank Journal of Physics: Conference Series, Volume 1065, Issue 4 Citation Wang Jintao *et al* 2018 *J. Phys.: Conf. Ser.* 1065 042029 Published under license by IOP Publishing Ltd

[19] 3D LASER SCAN. Startitaliana. http://www.startitaliana.com/areatest/prodotti_hscroll_laser.php#info

[20] 3D Laser Tank Calibration System. Tokheim ProGauge. https://www.pshglobalgroup.com/product/3d-laser-tank-calibration-system/

[21] ISO 17123-4:2012. Optics and optical instruments – Field procedures for testing geodetic and surveying instruments. Part 4: Electro-optical distance meter (EDM measurements to reflectors). Switzerland: ISO; 2012. p. 22

[22] Pollinger, F., Meyer, T., Beyer, J., Doloca, N. R., Schellin, W., Niemeier, W.,... & Meiners-Hagen, K. (2012). The upgraded PTB 600 m baseline: a high-accuracy reference for the calibration and the development of long distance measurement devices. Measurement Science and Technology, V. 23, No. 9, DOI:10.1088/0957-0233/23/9/094018, pp. 1-11

[23] OIML R 85-1&2:2008 Automatic level gauges for measuring the level of liquid in stationary storage tanks. Part 1: Metrological and technical requirements; Part 2: Metrological control and tests. Paris – France: Bureau International de Métrologie Légale; 2008. Available from: https://www.oiml.org. p. 49

[24] JCGM 100:2008 Evaluation of measurement data – Guide to the expression of uncertainty in measurement. Paris – France: Joint Committee for Guides in Metrology; 2008. Available from: https://www.bipm.org. p. 120

[25] JCGM 102:2011 Evaluation of measurement data – Supplement 2 to the "Guide to the expression of uncertainty in measurement" Extension to any number of output quantities. Paris – France: Joint Committee for Guides in Metrology; 2011. Available from: https://www.bipm.org. p. 72

[26] B. Delaunay: Sur la sphère vide, Izvestia Akademii Nauk SSSR, Otdelenie Matematicheskikh i Estestvennykh Nauk, 1934. pp. 793-800

[27] ISO/IEC 17025:2017 General requirements for the competence of testing and calibration laboratories. Switzerland: ISO/IEC; 2017. p. 30

[28] ISO/IEC 17043:2010 Conformity assessment — General requirements for proficiency testing. Switzerland: ISO/IEC; 2010. p. 39

[29] EA-2/14 M: 2008 Procedure for Regional Calibration ILCs In Support of the EA MLA. European Accreditation; 2008. p. 10

[30] ISO 13528:2015 Statistical methods for use in proficiency testing by interlaboratory comparisons. Switzerland: ISO; 2015. p. 57

Chapter 5

Properties of Tactile Sensors Based on Resistive Ink and the Dimension of Electrodes

Jaromír Volf, Viktor Novák, Vladimír Ryženko and Stanislava Papežová

Abstract

This chapter presents the ongoing research, which aims to select suitable electrodes for their use in the pressure distribution measurement system Plantograf. In our research, we examine more materials, especially Yokohama conductive rubber CS57-7RSC and also conductive inks, which are represented type DZT-3 K, Graphit 33 and mixture of Loctite-Henkel conductive inks type Loctite 7004Hr and Loctite NCI 7002EC. All materials can be used as a converter between pressure and electrical quantities in the design of planar pressure transduces. We build on our previous works, where were examined the properties of conductive rubber, conductive inks and electrodes. Next part is focused on the newest results of our research. Due to the still incomplete results in the given issue, we decided to perform an extensive and original measurement of a total of 172 combinations of different electrode sizes, the ratio of conductive ink mixtures and the thickness of the applied ink layer. Thanks to this, it will be possible in the future to select a suitable combination of electrodes and inks when designing tactile pressure sensors for industrial or medical applications without the need to perform time-consuming preparatory measurements and exclude unsuitable ink-electrode combinations.

Keywords: conductive elastomer, conductive ink, tactile sensor, foil transducer, pressure distribution

1. Introduction

Foil transducers are devices that convert the applied pressure into an electric signal. Under the applied pressure, the electrical resistivity changes as the material of the transducer are deformed. The pressure can be calculated by knowing the dependency curves of pressure and resistivity. The principle of the transducer is similar to a microphone: under the applied pressure, microscopic conductive particles approach is getting closer, which causes a decrease of the resistivity of the material. A basic comparison of the electrical and physical properties of individual materials of the transducer is presented within this contribution, focusing on thin foils of conductive ink and conductive rubber.

In the previous versions of the device PLANTOGRAF – that measure the pressure distribution on a surface – conductive elastomer Yokohama Rubber

CS57-7RSC was initially used; this material was commonly used in the production of tactile sensors [1–6]; different polymers and their properties are nearly described in [7–10]. However, this material with a thickness of 0,5 mm exhibits a relatively large hysteresis in the range of some percent, which prevents the measuring of the absolute pressure acting on the electrodes' system. The main problem with conductive rubber is the mutual influence of nearby electrodes, i.e., the currents flow horizontally between individual electrodes so that they affect each other. This could be avoided by the creation of a corresponding matrix consisting of individually separated sensors. This technology is however quite expensive and technically very demanding, given the required density of sensors. Therefore, we further experimented with conductive inks that exhibit only a very small mutual influence of nearby electrodes. Ink could be applied to the electrodes directly by offset print which is a considerably simpler method of application than separation of sensors. During our preliminary experiments, we found out that ink may not adhere to the electrodes fully so that alternative methods have to be investigated. The properties of individual transducer materials and possible methods of their use are presented in further paragraphs.

We subsequently focused on the use of conductive ink. Four types of conductive inks were obtained for the tests: NGAP FI Ag-4101 from the Spanish firm NANOGAP, Luxor from the Taiwanese firm Luxor, KH WS SWCNT from the Korean firm KH Chemicals, and DZT-3 K. The last type of ink was the only used in the measurements since owing to its composition; it could form a relatively high-quality conductive layer compared to the others. This ink uses carbon particles as filler. The other inks did not meet the requirements, either they were too thin and they did not form a continuous layer, or they did not adhere to the substrate (first two, both water-based inks) or they were excessively conductive – as the third ink with silver particles as a filler – the resistance of the ink was only in units of Ohm. Later we tested as conductive ink Graphit 33 spray and last attention was intended on Loctite - Henkel conductive ink Loctite-NCI 7002EC and Loctite-7004HR [4].

Now is under development a new measuring system Plantograf V20F. It is a flexible transducer, which can measure the skew surfaces.

2. Materials and methods

2.1 Design and principle of the transducer

The basic principle, cross-section and top view of the transducer Plantograf V16 are shown in **Figures 1** and **2**; the principle is the same both for the conductive ink and conductive rubber layer.

The current, represented by red arrows, flows from the inner electrode through the conductive rubber to the outer electrode. The voltage $U_x = 5$ V is displayed in the first figure. The common electrode, supplied by a voltage of 1.8 V and marked as "screening" is a common method used in printed circuits. It provides - by hardware - the separation of individual sensors and it prevents the mutual interaction of nearby sensors. The electrodes are etched onto a Cuflex film placed on the bottom of the sensor matrix. Cuflex is a thin Teflon foil with an applied copper layer, copper layer thickness 0.07 mm, foil thickness 0.09 mm, but several individual thicknesses of the Cuflex film are available. The surface of each electrode is completely covered. The electrodes and the conductive elastomer are protected from mechanical wear by a protective coating, a non-conductive flexible

DOI: http://dx.doi.org/10.5772/intechopen.101210

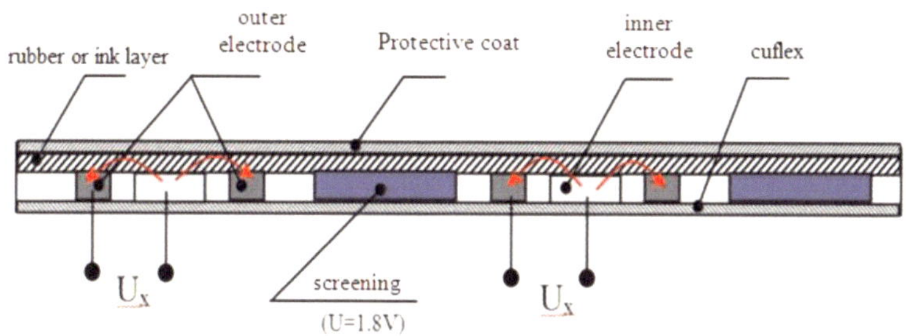

Figure 1.
Cross-section and top view of the transducer.

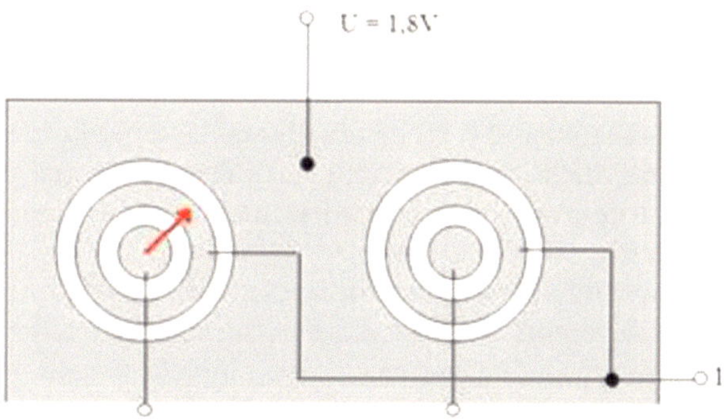

Figure 2.
Top view on the transducer.

material [5]. The resistance between the electrodes is measured, which is dependent on the deformation of the material under the pressure.

2.2 The Plantograf pressure measuring system and use of foil transducers

The foils are made of conductive ink or conductive rubber and are used in practice in the Plantograf measuring system. This is a tactile transducer which is picked up tactile information from a particular object and converts it into an electrical signal. The device is under steady development and improvement process of both the control electronic circuits and the transducer foil. Earlier versions of the device used conductive rubber, now we are still evaluating several conductive inks to meet all the required mechanical and electrical properties; thus, the use of the conductive ink is still rather experimental. First rubber-based sensors were used in 1978 wits USA made rubber Dynacon C for robot's hand. After 1982, the material was replaced with Yokohama Rubber with better mechanical and electrical. Plantograf, our transducer, was first designed in 1998. Since then, there were developed several versions of the transducer up to the current version Plantograf V16.

The system can process in real-time variable time pressure signals. It consists of 16,384 sensors (with a diameter of 2,5 mm each, in a 128 x 128 matrix arrangement) concentrated in the active area 0,5 x 0,5 m; it can sample and process up to 1000 fps. A full-frame is created by all 16,384 sensors. Plantograf measures the pressure distribution between the transducer and sole. Measured values are relative, where the analog output is converted on the digital signal by 8-bit A/D converter in 256 levels of gray. Every gray level was also assigned to pseudo colors. For exploitation full A/D convertor range is change get setting in range 0,5;1; 1,5; 2; 2,5; 3; 4; and

5. This way, the pressure distribution frame is represented in 256 color levels in a 2D or 3D model view. It is further possible to post-process the measured data via a dedicated program on a PC.

The system is used for dynamic and static measurements in various industrial or medical applications, described in more detail in [5, 11, 12]. Sample pictures of a graphic representation in the 2D and 3D view mode of the measured pressure distribution in a medical and industrial application of the device on a PC screen is shown in **Figures 3–5**.

2.3 Conductive rubber Yokohama CS57-7RSC

This transducer, made by Yokohama Rubber Co., labeled CS57-7RSC, represents a thin foil of conductive rubber (elastomer). This material is silicon-based and it is filled with small conductive pieces of pulverized graphite and as well as other additive particles (Iron (III) oxide and Silicon dioxide). The material is inert to water, methanol and ethanol and it is chemically stable. It exhibits relatively stable mechanical properties (thickness, elasticity) and values of electrical conductivity. The material has a tensile strength of 1.86 MPa, thickness of 0.5 mm, and a usable temperature range from −30–100°C [13].

These properties are fully appropriate for its use in the tactile sensor under normal operating conditions. A potential disadvantage of this material may be its low long-term stability of sensitivity and its known relatively high hysteresis. Hysteresis means the different values of the resistivity of the material in the process of loading and unloading under the same acting force, see also chart in **Figure 12**. These negative properties cannot be fully eliminated and they are given by the construction of the sensor and by the material itself [1–3]. A further issue is that rubber cannot be placed on individual electrodes separately or it would be technically extremely difficult by almost 16,500 electrodes.

Therefore, we started to experiment with various conductive inks, which do not exhibit these negative proprieties. Conductive ink can be applied to individual electrodes directly without major difficulties; this way, the mutual interaction of electrodes can be eliminated substantially as we obtain separated conductive places.

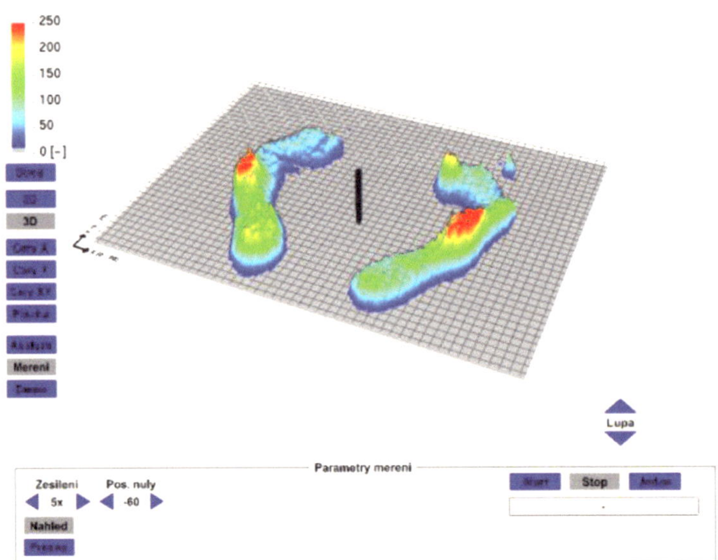

Figure 3.
Graphic representation of the distribution of pressure on a PC screen. Human foot in 3D.

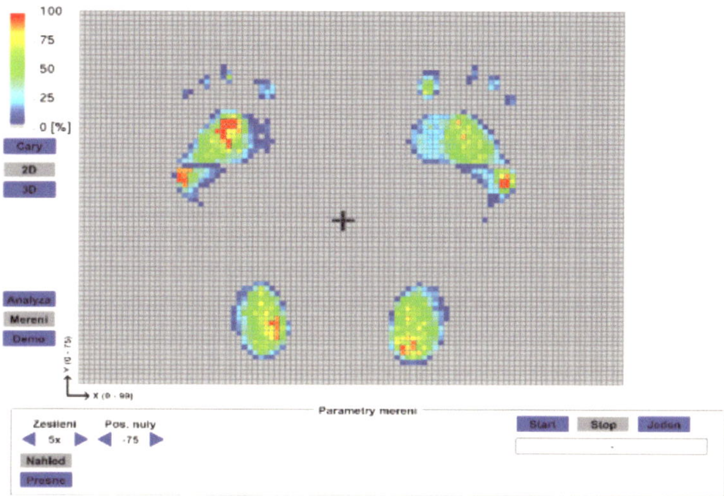

Figure 4.
Graphic representation of the distribution of pressure on a PC screen. Human foot in 2D.

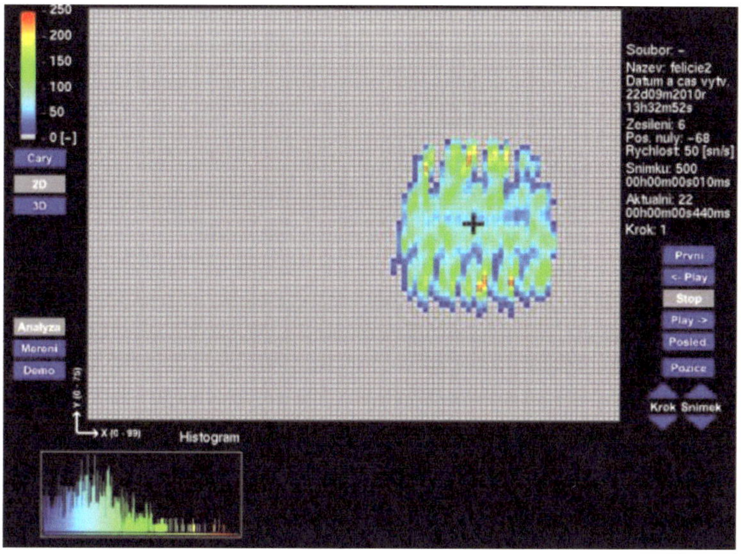

Figure 5.
Graphic representation of the distribution of pressure on a PC screen. Car Tyre.

2.4 Production of DZT-3 K ink specimens

Conductive ink is ink-filled with small pieces of conductive particles; we tested inks containing graphite and silver particles. For testing, we obtained four types of conductive inks: Luxor (Luxor, Taiwan), KH WS SWCNT (KH Chemicals, Korea), DZT-3 K (DZP Technologies, United Kingdom) and NGAP FI Ag-4101 (NANOGAP, Spain). The ink DZT-3 K was selected after preliminary evaluations and it was used in the measurements. This ink – using carbon particles as filler – could form a relatively high-quality conductive layer owing to its composition. The other inks did not meet the requirements: they were excessively conductive – as the third ink with silver particles as a filler – the resistance of the ink was only in units of Ohm. Other inks were too thin and they did not form a continuous layer or they did not adhere to the substrate (the first two - both water-based inks).

The electrical resistance of an unloaded ink layer should be optimally above 500 kΩ (at least 100 kΩ), and the resistance of the fully-loaded layer should be, optimally below 1 kΩ (usable range of hundreds of Ω). A possible disadvantage of the conductive ink compared with conductive rubber might be the difficulty of creating a compact and stable layer [5].

However, the selected ink DZT-3 K was unable to create a coherent conductive layer, i.e., to sustain its integrity when it was applied on the electrodes directly. Any negligible mechanical load caused the separation of the ink from the electrodes' surface. Additionally, we observed a certain deformation of the ink layer between the outer and inner electrodes. The measuring method – pushing with a force sensor tip on the ink layer – would not be applicable in this case.

We proceeded to an alternative procedure, as this setup proved not to be utilizable: the selected DZT-3 K ink was applied to the electrodes similarly as the conductive elastomer, by applying the ink on the surface of a 0,3 mm thick PET foil. TG 130 spray gun which can spray very low amounts of ink and enables fine control of spraying was used to deposit the ink on the foil. The ink DZT-3 K was diluted with water in the ratio of 1:1 to prevent the blocking of the jet. Three thicknesses of the deposited ink layer were selected: 7 μm, 15 μm and 23 μm. The spray applications were performed through a template made of the same foil with 3 mm holes given the 2.5 mm outer diameter of the circular electrodes for ink DZT-3 K. The thickness of the deposited ink layer was measured with a Mitutoyo SR44x1 digital micrometer with a measuring range of 0–25 mm and an accuracy of 1 μm. In a similar way was placed ink Graffiti 33 and Henkel ink, here was diameter to 3,5 mm.

2.5 Conductive ink Graphit 33

Graphit 33 is ink consisting of graphite particles, with organic solvent as a filler. The main properties of the ink are stated in the following **Table 1**.

This ink exhibited the same mechanical deficiencies as the previous specimen - poor adherence and incoherent conductive layer while applied to the electrodes directly, Therefore, we repeated the previous procedure and we deposited the ink on the surface of a 0,4 mm thick PET foil and it was applied to the electrodes similarly as rubber.

The ink was deposited on the foil in 20 μm, 50 μm and 80 μm layers (corresponding with 3x, 6x and 9x multiple spraying). There were used four types of electrodes for the measurement: LD, PD, LH, PH, the electrodes and their dimensions according to the following **Table 2**. The general design of the electrodes is depicted in **Figure 7**.

GRAPHIT 33	
Minimal full coverage layer thickness	**10–20μm**
Surface coverage	up to 4 m^2 / l
Packing	200 ml can
Drying time	<20 min
Color	black
Temperature limit of the graphite film	250–300°C
Temperature limit of the paint	±90°C
Surface resistivity	1000–2000 Ω

Table 1.
Properties of ink Graphit 33 [14].

Technology	Thermoplastic
Appearance	Black
Filler type	Carbon
Cure	Heat cure
Operating temperature maximum	100 °C
Product benefits	Screen printable Excellent screen residence time Flexible low-temperature drying cycles
Application	Conductive ink
Typical assembly applications	Force-sensitive modules, Printed resistors and Sensing devices
Key Substrates	Treated polyester and Polyimide
Emulsion thickness	20 to 40 μm
Viscosity, Brookfield, Speed 20 rpm, @ 20 °C	25,000 mPa·s (cP)
Density	1100 kg cm^{-2}
Shelf Life @ 5 to 30 °C (from date of manufacture)	365 days
Flash Point DIN 53213	78 °C

Table 2.
Properties of ink Loctite-7004HR [15].

2.6 Loctite - Henkel conductive ink Loctite-NCI 7002EC and Loctite-7004HR

We cooperate with the Faculty of Chemical Technology, University of Pardubice by creating test boards (see **Figure 6**) and developing of the new flexible transducer. These samples were made using printing technology. Each one of nine samples of electrode boards include 18 individual circular electrodes with different dimensions stated in **Table 3**; corresponding dimensions are graphically explained in **Figure 7**. Each test plate is created for the different mix of inks and thickness of ink.

Any of the nine samples test boards will have a different combination of the factors that may influence the sensitivity of the sensor, namely:

- thickness of the ink layer
- proportion of two conductive inks in the mixture
- dimension of the electrode

Before we have received the complete set of all nine test boards, we obtained a "pre-sample plate" to evaluate the design and the suitability for further measurements of the full set of samples. This "pre-sample" has ink thickness 25 (±1) μm and the proportion of the inks 60:40% in the mixture. The proportion of both conductive ink effects resultant resistance of the sensor. The ink Loctite-7004HR has by ink thickness 0,25 μm for loading state resistivity of about 3500 Ω/sq./mil. Properties of this ink are shown in **Table 2**.

The second used ink is Loctite-NCI 7002EC. This is ink with very high resistance. The ink Loctite-NCI 7002EC is non-conductive in the unloading state. Properties of this ink show **Table 4**.

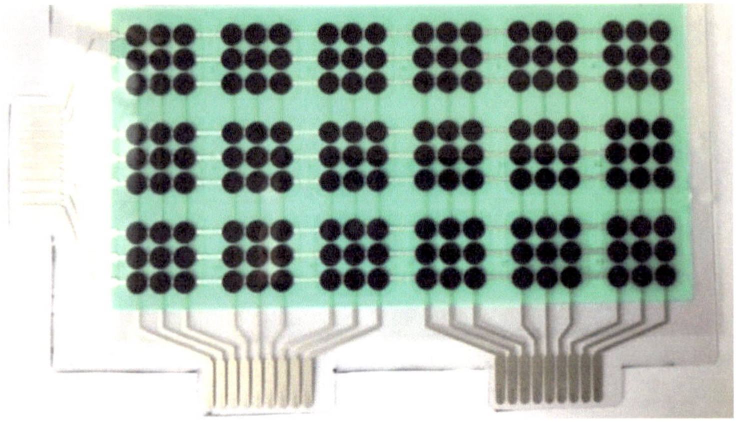

Figure 6.
Sample with circular electrodes.

Electrode number	R1	R2	R3	R4	E	M	S1	S2	Final S
1 - LH	0,2	0,45	0,55	1	2	0,1	0,5105	21,909	27,014
2 - PH	0,05	0,45	0,55	1	2	0,1	0,6282	21,909	28,191
3 - LD	0,2	0,5	0,75	1,25	2,5	0,25	0,6597	3141	38,007
4 - PD	0,05	0,5	0,75	1,25	2,5	0,25	0,7774	3141	39,184
5	0,05	0,825	0,925	1,75	3,5	0,1	2,13	69,318	90,618
6	0,05	0,8	0,95	1,75	3,5	0,15	20,024	67,846	8787
7	0,05	0,775	0,975	1,75	3,5	0,2	18,788	66,334	85,122
8	0,05	0,75	1	1,75	3,5	0,25	1759	64,784	82,374
9	0,05	0,725	1025	1,75	3,5	0,3	16,432	63,193	79,625
10	0,05	0,7	1,05	1,75	3,5	0,35	15,313	61,564	76,877
11	0,05	0,675	1075	1,75	3,5	0,4	14,233	59,895	74,128
12	0,05	0,75	0,85	1,6	3,2	0,1	1759	57,716	75,306
13	0,05	0,725	0,875	1,6	3,2	0,15	16,432	56,362	72,794
14	0,05	0,7	0,9	1,6	3,2	0,2	15,313	54,968	70,281
15	0,05	0,675	0,925	1,6	3,2	0,25	14,233	53,535	67,768
16	0,05	0,65	0,95	1,6	3,2	0,3	13,193	52,063	65,256
17	0,05	0,625	0,975	1,6	3,2	0,35	12,192	50,551	62,743
18	0,05	0,6	1	1,6	3,2	0,4	1123	4,9	6023

Table 3.
Parameters of electrodes.

2.7 Shape of the measured electrodes

The measurements of individual electrodes were performed on a scanning matrix comprising circular electrodes with a 2–3,5 mm diameter; the electrodes were placed on an elastic printed circuit board. The same sizes and design of the electrodes as described were used to enable the comparison of the properties of tactile sensors with a conductive elastomer with those with a conductive ink [5]. Conductors were soldered to the outlets of lines and columns which enabled the easy choice of a particular electrode. The dimensions of the measured electrodes are presented in **Figure 7**.

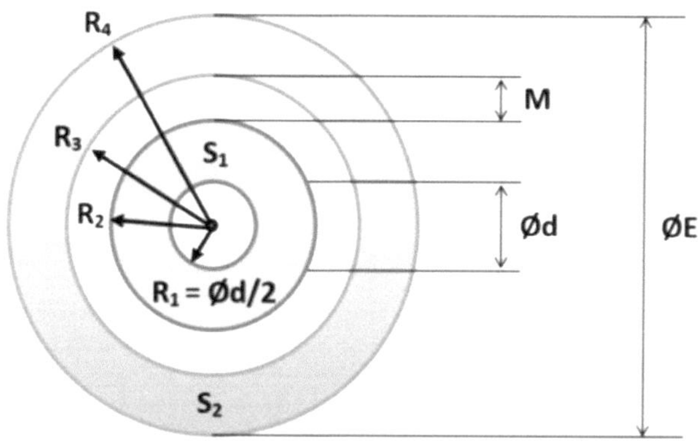

Figure 7.
Dimensions of the measured electrodes.

Technology	**Thermoplastic**
Appearance	Black
Cure	Heat cure
Operating temperature maximum	100 °C
Product benefits	Non-conductive Screen printable Flexible resistive Excellent screen residence time Flexible low-temperature drying cycles Good adhesion
Application	Non-Conductive ink
Typical assembly applications	Printed resistors Sensing devices Heating elements Protection against electrostatic discharge (ESD)
Key Substrates	Treated polyester and Polyimide
Solid's content	37.5 to 40.0%
Viscosity, Brookfield, Speed 20 rpm, @ 20°C	10,000 to 25,000 mPa·s (cP)
Density	1270 kg cm^{-3}
Theoretical coverage, @ 10 μm dry coating thickness	14 m^2 kg^{-1}
Shelf Life @ 5 to 30°C (from date of qualification in original seal)	365 days
Flash Point DIN 53213	78 °C

Table 4.
Properties of ink Loctite-NCI 7002EC [16].

The electrodes are denoted accordingly to their marking: S1 – inside electrodes area, S2 – outside electrodes area, S – both electrodes area. The parameters of the electrodes are listed in **Table 3**, dimensions are stated in mm.

2.8 Determination of electrical resistance of the ink

The measurement circuit diagram is depicted in **Figure 8**. It represents a stabilized circuit, that supplies a voltage divider. One resistor of the divider is constant

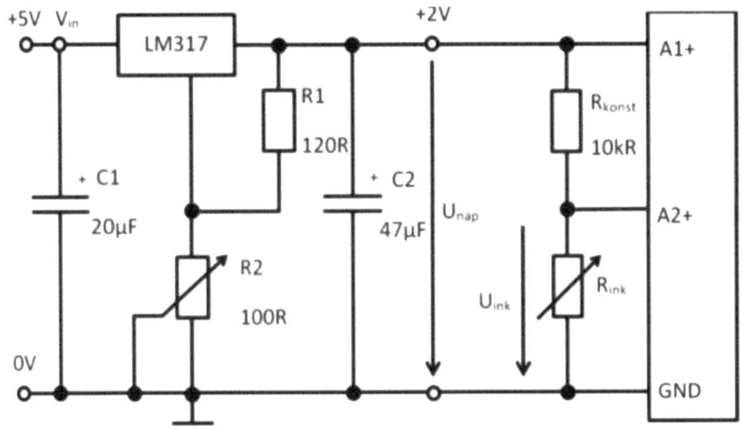

Figure 8.
Circuit diagram for measuring the conductive ink resistance.

and the other one is variable, represented by the resistance of the conductive ink. The supply circuit, is LM317 voltage stabilizer, enables the setting of the supply voltage to 2 V and its fine adjustment. The low value of the supply voltage ensures that only a small current flow through the circuit, and thus it avoids the heat of the conductive ink. The electrical resistance of the constant resistor in the divider is 10 kΩ, to ensure a constant current in the divider circuit. The electrical resistance of the ink was calculated using the formula (1):

$$R_{ink} = \frac{R_{konst} . U_{ink}}{U_{nap} - U_{ink}} \quad (1)$$

where
R_{ink} resistance value of sensor
R_{konst} resistor with constant resistance value 10 kΩ
U_{ink} voltage on sensor
U_{nap} power voltage

The measurement of the voltage on the conductive ink – needed to calculate the resistance values in the divider – was determined by the measuring card NI 6008. The voltage UINK was connected to an analog input of the card and it was measured by the RSE method (Reference Single Ended) against ground potential. On the other analog input of the card was connected to the output voltage UNAP from the stabilizer in the same way. The output of the measuring card was connected to a PC via USB. The entire measuring station was controlled by the NI LabView program. A LabView application was also created, which enables the recording and the calculation of the electrical resistance of the conductive ink [5].

2.9 Measurement procedure

Measurements of the properties of conductive ink were performed at a robotized workplace equipped with a Turbo Scara SR60 robot. The basic step of the vertical motion of the robot's arm is 0.01 mm. The pressure was applied by the measuring tip 3 mm in diameter using the vertical motion of the robot's arm. The arm was moved in 0.02 mm increments for a general overview of the behavior of an electrode and further in 0.01 mm step for a more detailed analysis. The pressure was applied using the vertical motion of the robot's arm, at which the Hottinger DF2S-3 tensometer force sensor was fixed. The measuring tip is a circular surface with ϕ3 mm or ϕ4 mm diameter. The output voltage of the type DF2S-3 tensometer force sensor was measured by an Almemo 2890–9 Data Logger.

This force sensor was chosen because of its appropriate range and high sensitivity. Its accuracy is 0,03%, nonlinearity is 0,03%, max. Loading is 200 N, and sensitivity 2 mV/V, power supply 5 V, see [17]. For connecting to data logger was used connector Ahlborn Almemo ZA9105650FS1V with input range ± 2,6 V, resolution 1 μV, precision class AA. Detailed technical parameters see in [18]. This way the electrical resistance of the conductive ink was measured.

The control unit is set up to display the values in grams; the conversion into the pressure values was made subsequently. The foil with the deposited ink was placed on the electrode field. The measuring tip touched down on the surface of one tactile point and pressed on the conductive ink deposited on the foil against the circular electrodes. The pressure imposed on the electrodes was calculated from actually exerted force and the known area of the surface of the measuring tip. **Figure 10** presents the overall layout of the measuring post. **Figure 9** shows the detailed view on the measuring head: (1) is for the conductive ink deposited on a foil, (2) indicates the measuring tip, (3) is the force sensor DF2S-3 and (4) indicates the robot's head.

The movement of the robot's arm was controlled by a dedicated remote control. The foil with deposited ink layer and the tactile field with circular electrodes was put into the workspace of the robot's arm and it was placed on a hard, plastic pad. The positioning of the measuring tip was performed by manual control of the robot. After checking that the measuring tip covers the entire surface of the electrodes, the Almemo 2890–9 data logger was reset to zero in an unloaded condition. For a correct measuring, the stable "unloaded" resistance value of the ink layer, depending on the thickness of the layer and the electrodes' design, was determined before further loading of the electrodes. This value and the matching vertical coordinate were recorded as a start value. Subsequently, the robot's arm was gradually lowered using the + and – buttons on the remote control.

After a couple of test measurements, the maximum vertical shift of 180 μm was determined, representing 18 down steps of the robot's arm. That was sufficient to measure the boundaries when loading force does not more decrease the resistance of the ink. From this value, the backward (up) movement by the same number of steps was carried out (unloading). Loading and unloading procedures were performed to measure the hysteresis of the conductive ink. Corresponding resistance and pressure

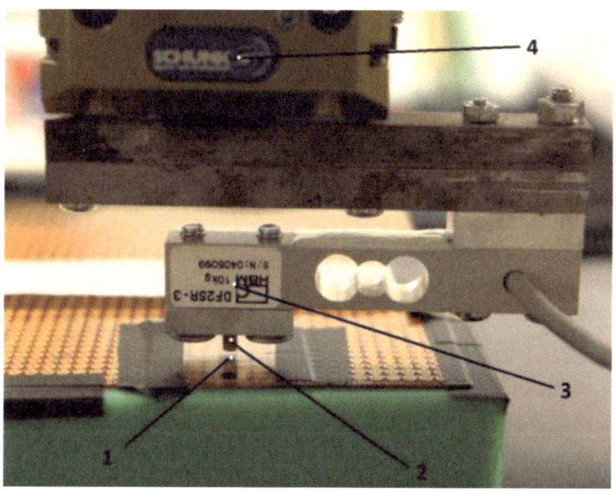

Figure 9.
Detailed view on the measuring head.

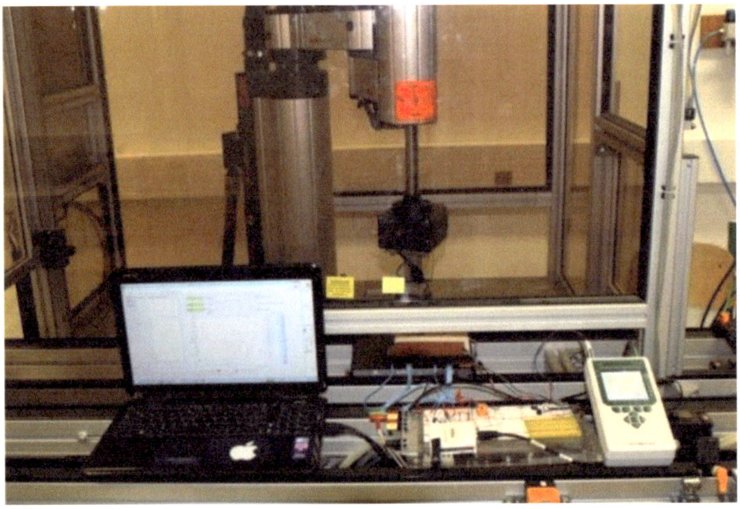

Figure 10.
The layout of the robotized measuring post.

values were logged by the Lab-View application after each shift. For measurement, we used LabView measure card NI USB 6009 with the common GND and with inside A/D convertor 13 bits and input range 5 V [4, 19]. One measurement cycle thus contained 37 values. Between the individual measurement cycles, there was a five-minute break, so that the material could relax unloaded. The measurements were repeated 10 times for all types of electrodes and each ink layer thickness.

Loctite inks were measured via LabView card full automatically using a robot control program. The full set consists of 172 samples of combination electrode dimension, a mix of inks and thickness of ink layers. The voltage on sensors was measured by LabView card. Were calculated average value, standard deviation and then standard uncertainties – type A, B and combination standard uncertainties type C. These uncertainties are marked for lucidity only in separate graphs of dependencies. This same process was used for conductive ink DZT-3 K, too. For Yokohama, conductive rubber uncertainties were not measured. In lucid graph with more curves was used only average values.

3. Results

3.1 Results with conductive elastomer Yokohama CS57-7RSC

The measurements were carried out ten times on the same sensor point for conductive elastomer and each type of electrode, i.e., both LD and PD. The diagrams below this paragraph show the dependence of the measured electrical resistance on the applied pressure, as an average from each of the ten sets of measurements. The electrical resistance should decrease due to the applied pressure, based on the principle explained at the beginning of the chapter. Now the course of the electrical resistance for selected combinations of materials and electrodes must be studied and their appropriateness for the described device assessed.

The first diagram in **Figure 11** presents the dependence of electrical resistance on the applied pressure for conductive Yokohama CS57-7RSC. In the diagram, the behavior of the dependence for a PD electrode by the red curve and for an LD electrode is given by the green curve.

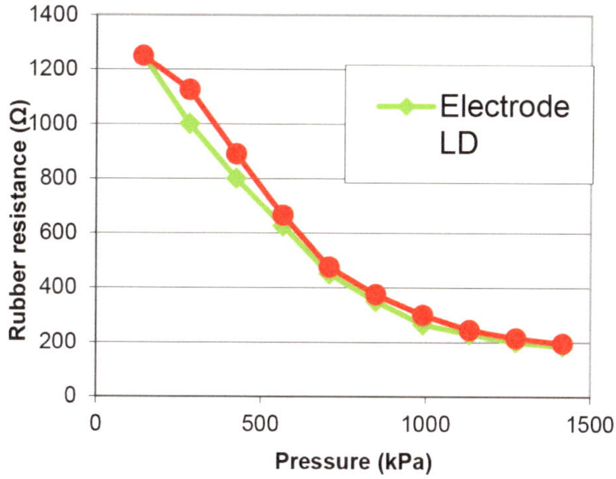

Figure 11.
Dependence of electrical resistance on the applied pressure – Conductive rubber, LD and PD electrodes.

Yokohama elastomer was tested on electrodes 1–4, see **Table 3**. The resistivity changes between 200 and 1300 Ω by pressure between 100 and 1400 kPa.

Further, the hysteresis curve of elastic material Yokohama was measured. Hysteresis curves measured by loading (red curve) and unloading (green curve) cycle shows **Figure 12**.

3.2 Results with conductive ink DZT-3 K

The measured data for LD-type and PD-type electrodes for all 3 thicknesses of the ink layers - 7 μm, 15 μm and 23 μm – are represented graphically in the following graphs. All measurements were repeated 10x and the total (combined) measurement uncertainty was calculated and graphically represented by respective intervals for each measured value.

Figures 13 and **14** present the graphical result for the conductive ink of thickness 7 μm in the loading and unloading cycle for LD electrode types. Initial insensitivity is

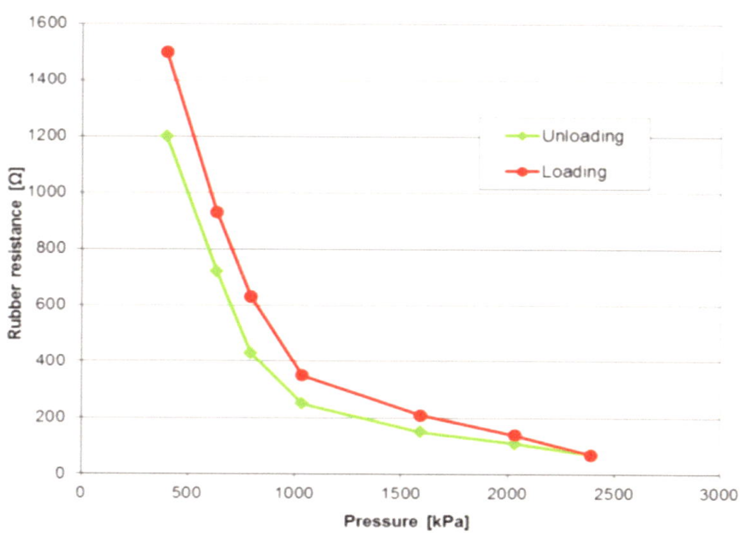

Figure 12.
Hysteresis of the conductive elastomer.

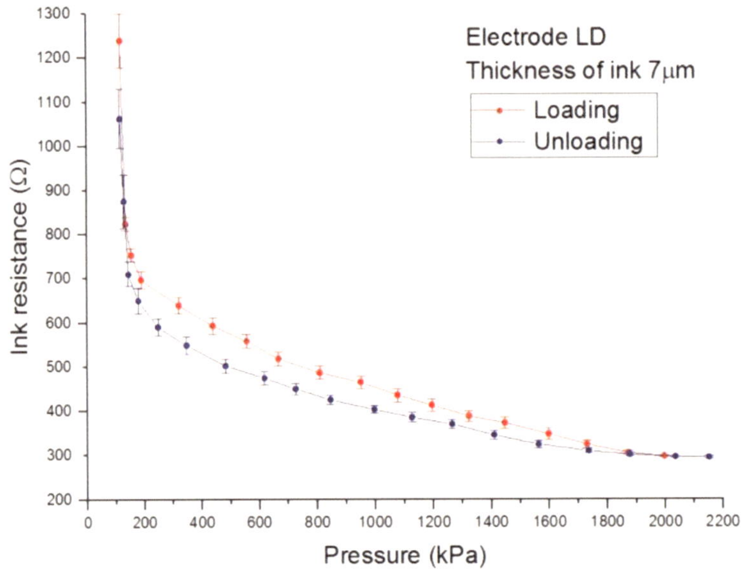

Figure 13.
Dependency of the resistance of a 7 μm ink layer in the loading and unloading cycle of LD electrode.

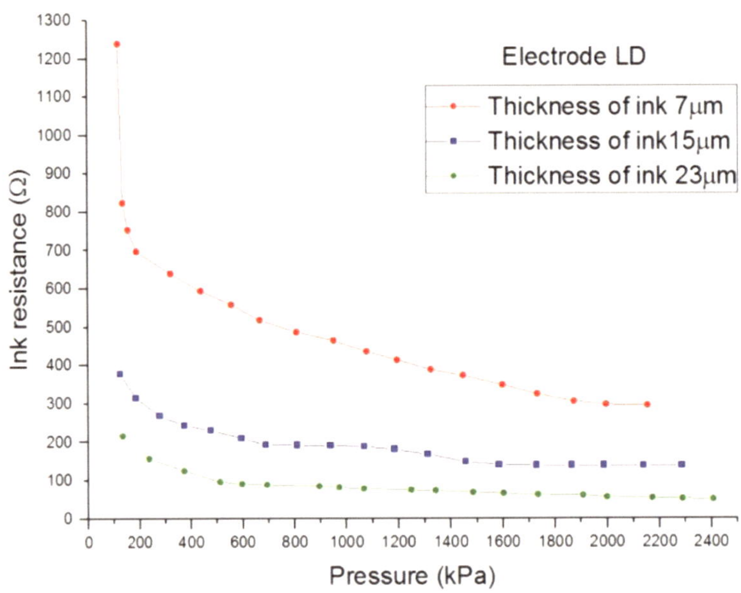

Figure 14.
Comparison of LD-type electrodes.

apparent, it may be caused by the force necessary for the touch-down of the foil with the deposited ink on the electrodes. Hysteresis is apparent in all electrodes, similarly as in the case when a conductive elastic material was used, however, it is much lower.

Figures 15 and **16** give the comparison of the courses of electrical resistance during loading of LD-type and PD-type electrodes for various thicknesses of the deposited ink layer. From the diagrams, it is apparent that maximum sensitivity is achieved for a 7 μm thickness of the deposited ink layer for both types of electrodes.

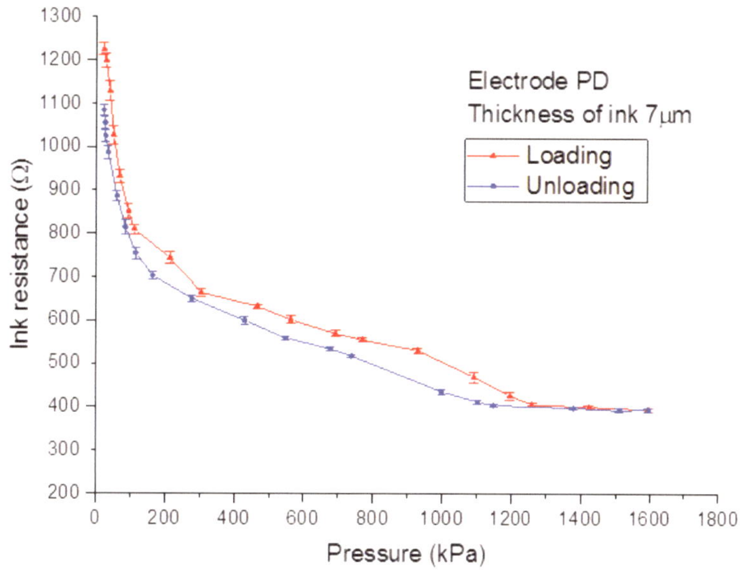

Figure 15.
Dependency of the resistance of a 7 μm ink layer in the loading and unloading cycle of PD electrode.

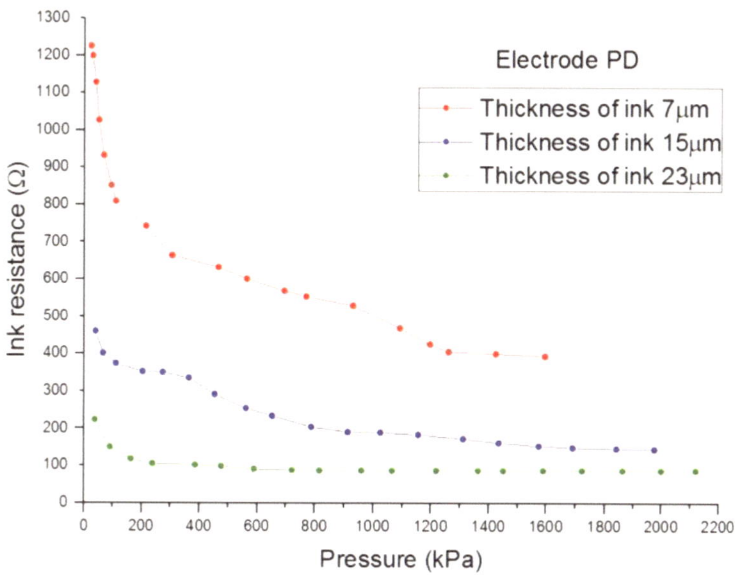

Figure 16.
Comparison of PD-type electrodes.

The last graph in **Figure 17** presents the overall results of the measurements; the graph shows a comparison of changes in resistance depending on the pressure for the LD and PD-type electrodes for different thicknesses of the applied conductive ink. The greatest sensitivity is achieved for a 7 μm ink layer. The comparison of LD and PD electrode types shows a somewhat smoother curve course of the PD electrode for the ink layers of 15 μm and 23 μm. Contrarily, the curve is smoother for the LD electrode and the ink layer of 7 μm.

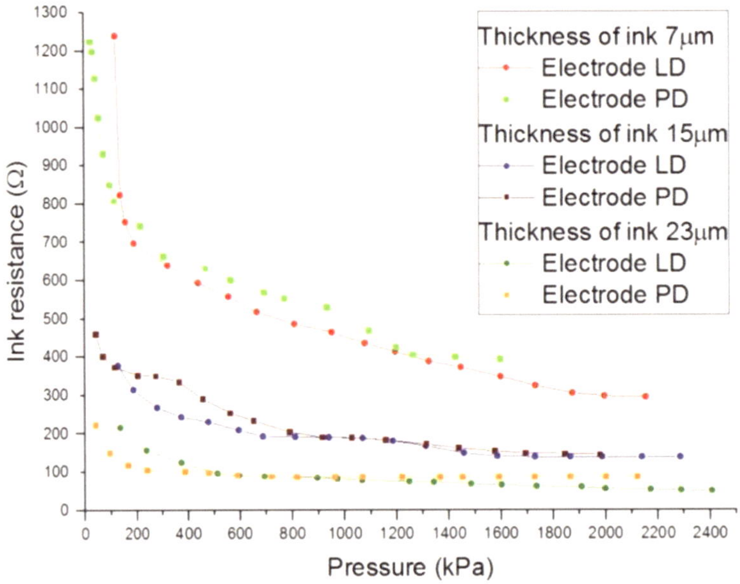

Figure 17.
Overall comparison of LD and PD electrode types.

3.3 Results with Loctite - Henkel conductive ink Loctite -NCI 7002EC and Loctite-7004HR

The loading force was exerted from 0,37 N up to cca. 17,6 N. This resulted in the measured range of pressure values approx. From 30 kPa up to 1400 kPa for the particular measuring tip.

For measuring these inks, we prepared nine test desks (**Figure 6**), each with the 18 electrodes types, see **Table 3** and **Figure** 7. Each test desk was prepared with a different mix of inks (40:60, 50:50 and 60:40) and three different ink layer thicknesses (10 μm, 20 μm, 30 μm).

The presented results were measured on the test desk Nr. 10, where mix of inks ECI7004HR:NCI7002 is 60:40% and the thickness of the ink is 20 μm. The initial loading pressure varies significantly, as every electrode has own threshold, when it starts to react to the applied pressure (i.e., when the resistance starts to drop). The scale on the graphs has been maintained the same to enable visual comparison of the courses. The graph on **Figure 18** shows the electrode 3 - LD and **Figure 19** show the electrode 4 - PD.

The course of the electrode LD is very particular compared to other electrodes. It has a very steep beginning part and later the resistance is almost linear.

The course of electrode 4 has a similar course as electrode 3, but the run part is more sheer. **Figure 20** show resistance dependence on pressure for both electrodes with uncertainties.

The next graphs show resistance dependence on pressure for all electrodes 1 to 18. It represents the most extensive measurements, which we ever made. Fully we made about 172 measurements of different full courses for different 18 electrodes, mix conductive inks and different thicknesses of these inks. Every course was measured 10 times, it presents more than 1700 number measurements full graphs. The way of measurement is described in Chapters 2.8 and 2.9. The comparison of all individual electrodes is shown in following **Figures 21** and **22**.

From graphs appear that some electrodes are suitable for low loading, e.g., electrodes 5, 7, 8, 9, 11 and 13, see **Figures 21** and **22**. Electrode 5, see **Figure 15**, exhibits

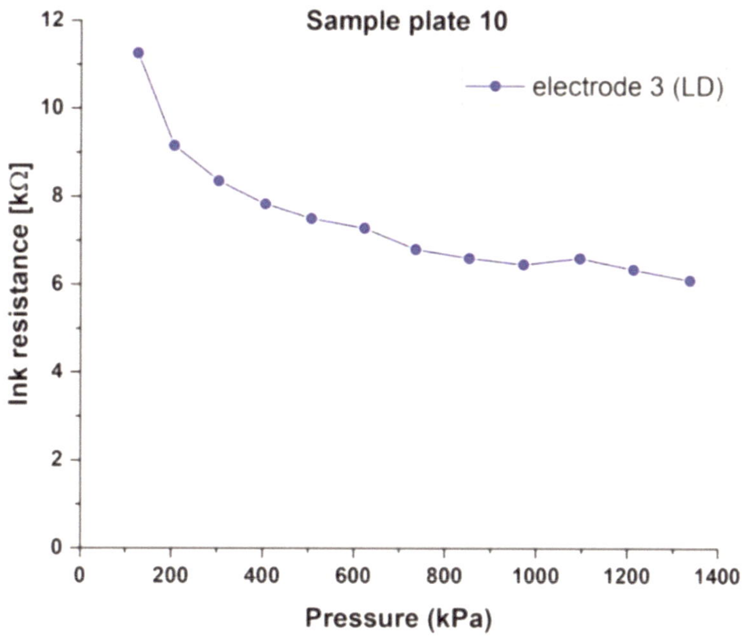

Figure 18.
Electrode 3 – LD.

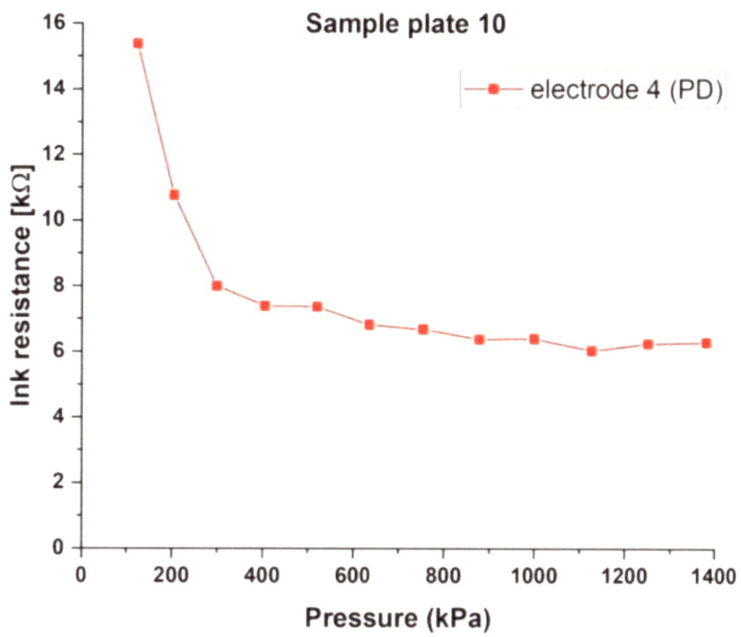

Figure 19.
Electrode 4 – PD.

a typical course as demonstrated by most of the electrodes – initial steep descent in the resistance, turn and following little loss in resistance with stagnation towards the end. Electrode 10 has a gradual decrease in resistance up to cca. 400 kPa, followed by stagnation. Electrode 12 exhibits a typical course; notable is that its construction enabled the measurement starting at cca. 50 kPa. Electrode 15 exhibits similar behavior as electrode 10, the measurement is loaded with significantly

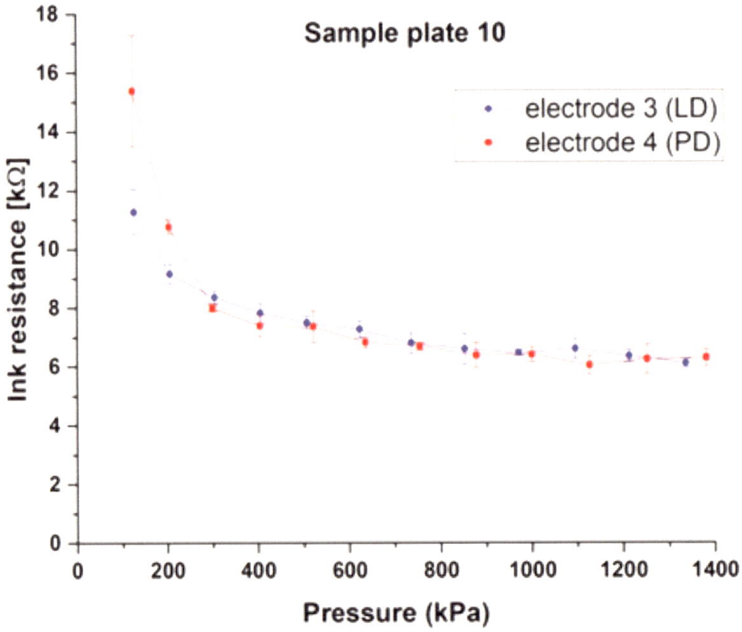

Figure 20.
Resistance dependence on pressure for both electrodes LD and PD with uncertainties.

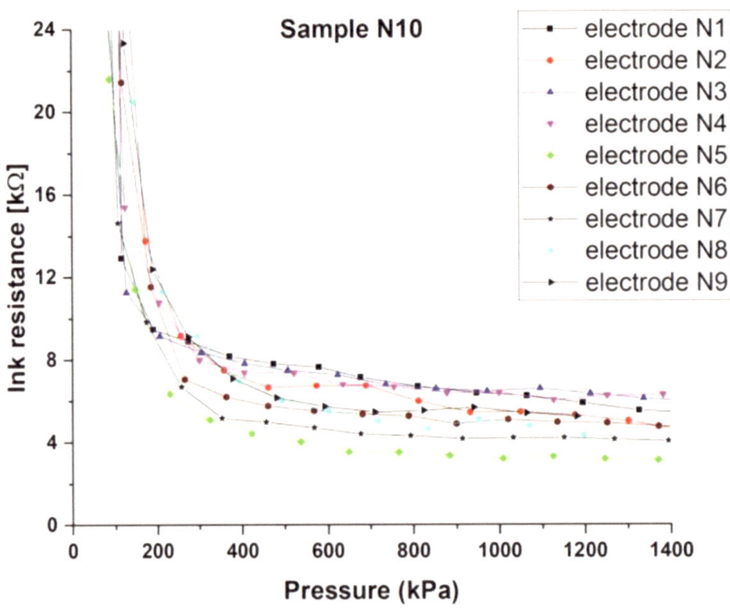

Figure 21.
Dependence of resistance on pressure for electrodes 1–9.

less uncertainty and starts at lower pressures. Electrode 14 has a typical course; its dimensions do not differ significantly to others. The courses are convenient for measuring the lower pressure range. Also, note the similar dimensions of both electrodes 16 and 17.

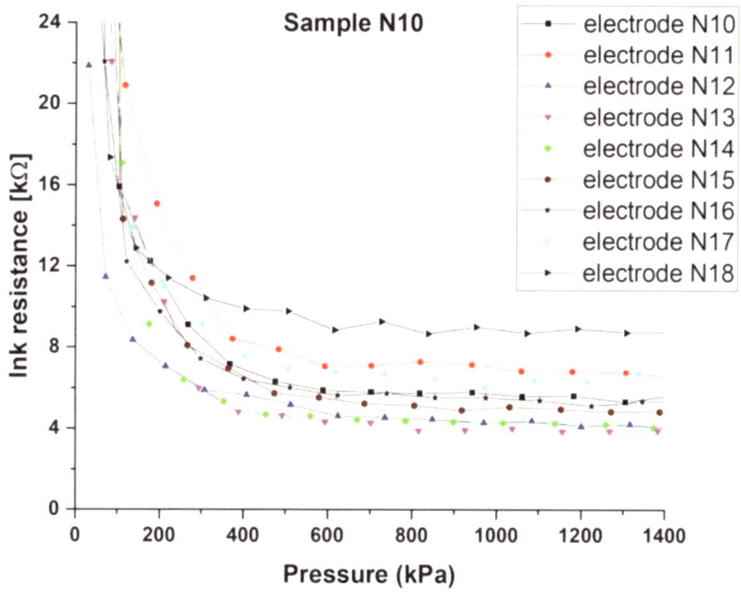

Figure 22.
Dependence of resistance on pressure for electrodes 10–18.

4. Discussion

In this chapter, we will discuss the comparison of the conductive ink and elastomer material, various electrode types and respective ink layer thicknesses. Generally, it may be stated that the measured electrical resistance of the material depends on the applied force and the contact area of the object with the transducer. The main issue is the mutual interaction between nearby sensors which is significant mainly in the rubber material.

Yokohama conductive rubber CS57-7RSC was loaded with a lower maximum pressure than ink according to the manufacturer's technical specification. Higher pressures may decrease the accuracy of the measurement and may destroy the rubber. The usability of this material in the Plantograf planar transducer was confirmed with good results within a relevant data output for both types of electrodes in the pressure ranges from 200 kPa up to 1400 kPa. From loads of approx. 1000 kPa the slope of the curve decreases which causes a lower sensitivity of the transducer for higher pressures. Due to mutual interaction of the nearby electrodes, it cannot be used to measure the absolute value of the pressure on the surface directly but only the pressure in a relative scale, e.g., 0–255 kPa on an 8-bit converter like the Plantograf can be determined. The absolute value of the pressure over every electrode may be determined using transcendental equations.

After preliminary measurements with different types of electrodes we found out that the electrical resistance is influenced mainly by the space between the inner and outer electrode and by the size of the electrode itself – see **Figure 7**. However, PD and LD electrodes differ only in the diameter of the inner opening – i.e., 0.1 mm (PD) and 0.4 mm (LD). Due to their similar parameters, the difference in the electrical resistance of these two electrodes is therefore not significant; only the behavior of the dependence slightly varies, as can be seen in **Figures 14** and **16**. A thicker ink layer contains so many carbon particles that after applying pressure on the ink layer, its behavior is approaching that of a conductor with an insignificant resistivity change.

Mutual interaction with conductive ink DZT-3 K is very low (max. up to 3%), demonstrated by measuring on the 5x5 and 3x3 sensor matrix. At loads above 600 kPa with thicker ink layers (here 15 μm and 23 μm) there is only a slight change of the electrical resistivity. Therefore, these configurations are not appropriate for higher loads. The best sensitivity was achieved with the 7 μm ink layer. The curve of the PD-type electrode is less smooth than that of the electrode LD, with a nearly linear dependence in the range from 200 kPa to 2000 kPa. The setup with an LD electrode and a 7 μm ink layer was assessed as the best transducer in the experiment [20].

DZT-3 K ink showed significantly higher sensitivity than Graphit 33 ink, when it reacted to a change in resistance at a lower pressure around 22 kPa. Graphit 33 reacted at a pressure of 267 kPa. For this comparison, it is necessary to add that the measured thinnest layer of DZT-3 K ink had half the thickness of the thinnest layer measured in this work. Even with a larger layer thickness, Graphit 33 showed a higher resolution. The resistance for the thinnest layer varied from 3.4 kΩ to 380 Ω and the other two thicknesses had a lower but still acceptable resolution. This fact was stated by the author in his work, i.e., that the DZT-3 K ink has high conductivity and thus a low resolution. When measured on the thinnest layer, its resistance ranged between 1.2 kΩ and 390 Ω. For wider layers, the resistance varied minimally.

In general, every electrode exhibits an initial steep decrease in the electrical resistance followed by a turn, when the resistance decreases significantly more slowly with the rising pressure. This turn is situated in the pressure range 200–400 kPa, depending on the electrode, also the turn is differently sharp. This phenomenon is caused due to the exponential dependency of the resistance on the pressure, which bases on the composition of the material; as the pressure is high enough, there are created significantly fewer conductive paths, thus the resistance drops only a little.

Further, the uncertainties are generally much greater in the range of lower pressures, particularly in pressure ranges under the described turn. These are given first by the light contact of the measuring tip with the surface and secondly by the light contact of the conductive layers itself with the electrode.

Most electrodes exhibit also stagnation in the loss of electrical resistance towards high-pressure levels. This is caused due to the saturation of the material as the particles are compressed to their maximum so the electrical resistance cannot drop anymore. The usable range of the electrodes is therefore limited to pressure ranges below the saturation level.

The particularity of electrode 1 was probably caused due to its different dimensions compared to other electrodes. After verifying the course with a more detailed measurement, it may be a suitable electrode for measuring higher pressure ranges due to its almost linear characteristics and better sensitivity in the higher-pressure ranges.

From the courses of the resistance, it can be seen, that the electrodes are not "universal", i.e., usable in the whole pressure range (with the exception of the electrode 1). Also, the upper value of pressure is limited to cca 500 kPa, then the drop in resistance is negligible. However, this poses no problem, as such high pressures are not expected to be measured in common industrial or agricultural applications.

5. Conclusion

Based on our experiments, we conclude that both materials – conductive ink and conductive rubber – are suitable as transducers of pressure and electrical quantities, however, with some limitations. Conductive rubber does not meet the requirements of the Plantograf, as it has a limited pressure range for higher loads. Furthermore, there is significant hysteresis of this material, which appeared during the loading and unloading of the rubber. It is a limiting factor for its use in newer versions of the Plantograf measuring system, which are capable of the real-time measurement of applied pressure up to 1000 frames per second. This phenomenon also prevents the measuring of the absolute pressure acting on the electrode system.

Conductive ink exhibited better results than conductive rubber, namely in the setup with the LD electrode and the 7 μm ink layer. However, there is another problem: the used ink can be wiped off the electrodes very easily. This water-based ink sinks into the gaps between electrodes and it and exhibits small adherence to the electrodes. Thus, we had to select an alternative procedure by spraying the ink on the foil. To allow direct application of the ink to the electrodes, different, polymer-based types of ink may get better mechanical properties. Inks with polymer-based binders adhere to the electrodes much better and the layer is not excessively destructed by the applied pressure [5, 13, 21, 22].

From a comparison of the properties of two types of conductive inks DZT-3 K and Graphit 33, it can be concluded that neither of them is the most suitable choice for tactile sensors. Each would be suitable in a different application where either greater or lesser sensitivity would be required. It is possible to take into account the fact that if the ink with graphite particles were measured on the same thin layer as the second type of ink, it could show greater sensitivity. In addition, the life of the ink layer that will be required when used in a tactile sensor needs to be taken into account. When measuring ink with carbon particles, there is no mention of damage to the surface of the layer. In the measurement of ink with graphite particles, there were cases of a deformed layer. It is, therefore, necessary to consider the cause of the deformation; whether the layer is poorly applied or, for example, the load pressure is too high.

Based on the performed preliminary measurement of the sample plate, we can conclude that the measuring methodology and computer processing of the data is adequate, however, to measure the full set of the samples, there have to be done some minor adjustments. First, more focus is to be given to lower pressure ranges up to cca. 500 kPa, hence the electrical resistance does not change significantly with higher pressures, given the saturation of the material. For more detailed measurement, a smaller step (0,01 mm) may be considered for some electrodes to determine the course more accurately, this applies particularly for low-pressure loads (below 200 kPa). The control program of the robot will have to be adjusted accordingly. The dimensions of the electrodes have a partial impact on the course of the resistance-pressure curve, which is mainly demonstrated with electrode 1, which has significantly different dimensions than others. This electrode is also the most suitable for measuring higher pressures up to 1000 kPa. Other electrodes have their working range up to cca 500 kPa, which is sufficient for their proposed applications.

Generally, the range of the pressures that can be measured using both conductive rubber (earlier research) and various setups with conductive ink (our current research) is from tenths of kPa up to cca 2000 kPa. There are visible some significant dependencies (electrode size, thickness of ink layer, mixture composition) on the sensitivity and applicable pressure range. We preliminary found out, that for instance thinner ink layer causes significantly higher sensitivity of the transducer; the size of the gap between the inner and outer ring of the electrode extends the measurable pressure range; the drop of electrical resistance is less steep with a growing gap. However, these factors are still subject to ongoing research and statistical evaluation, final results will be available probably within 2 years.

In any case, the new design of the electrodes proved to be capable for the proposed use in foil transducers between pressure and electrical resistance, the main concern is now the usable pressure range. In further measurements, other significant dependencies may be discovered, namely the impact of the thickness of the ink layer and the ratio of the ink mixture on the sensitivity and usable range of the electrodes.

Acknowledgements

Supported by the Technology Agency of the Czech Republic, Czech Republic, Project No. FW01010217.

Author details

Jaromír Volf*, Viktor Novák, Vladimír Ryženko and Stanislava Papežová
Faculty of Engineering, Department Electrical Engineering and Automation, Czech University of Life Sciences, Prague, Czech Republic

*Address all correspondence to: volf@tf.czu.cz

References

[1] Barman S, Guha SK. Analysis of a new combined stretch and pressure sensor for internal nodule palpation. Sensors and Actuators A-Physical. 2006;**125**(2):210-216

[2] Souza FG, Michel RC, Soares BG. A methodology for studying the dependence of electrical resistivity with pressure in conducting composites. Polymer Testing. 2005;**24**(8):998-1004

[3] Soares BG, Amorim GS, Souza FG, et al. The in situ polymerization of aniline in nitrile rubber. Synthetic Metals. 2006;**156**(2-4):91-98

[4] Volf J, Novák V, Ryzhenko V. Effect of conductive ink properties of tactile sensors. Procedia Engineering. Elsevier. 2015;**120**:200-205

[5] Volf J, Novak D, Novak V, Papezova S. Evaluation of foil transducers and their use in tactile sensors. Measurement. 2019;**136**:573-578

[6] Koder P, Novak V, Ryzhenko V, Hruby D, Volf J, Novak D. Plantograf V18- New construction and properties. Agronomy Research. 2018;**16** (Special issue 1):1085-1094

[7] Giovanelli D, Farella E. Force sensing resistor and evaluation of technology for wearable body pressure sensing. Journal of Sensors. Hindawi. 2016:9391850

[8] Seo S, Kim S, Jung J, Ma R, Baik S, Moon H. Flexible touch sensors made of two layers of printed conductive flexible adhesives. Sensors. 2016;**16**(9):1515

[9] Maddipatla D, Narakathu B, Ali M, Chlaihawi A, Atashbar M. Development of a novel carbon nanotube based printed and flexible pressure sensor. In: 12th IEEE Sensors Applications Symposium, SAS. 2017. p. 7894034

[10] Gao L, Zhu C, Li L, Zhang C, Liu J, Yu H, et al. All paper-based flexible and wearable piezoresistive pressure sensor. ACS Applied Materials and Interfaces. 2019;**11**(28):25034-25042

[11] Prikner P, Volf J. Contact pressure distribution in tyre tread pattern. In: Proceeding of 6th International Conference on Trends in Agricultural Engineering. Prague, Czech Republic; 2016. pp. 522-527

[12] Volf J, Vlček J, Holý S, et al. Measurement in Biomechanics by using of Area Pressure sensor. In: 6th World Multi-Conference on Systemics, Cybernetics and Informatics (SCI 2002)/8th International Conference on Information Systems Analysis and Synthesis (ISAS 2002). Orlando, FL USA; 2009. pp. 276-281

[13] Technical documentation of the conductive composite elastomer CS 57-7 RSC. Japan: Yokohama Rubber Co. Ltd; 1980

[14] Available from: https://static.rapidonline.com/pdf/87-0695.pdf, [Accessed 17 September 2021]

[15] Available from: https://www.eptanova.com/sites/default/files/download_products/loctite-eci-7004hr-ec.pdf, [Accessed 17 September 2021]

[16] Available from: http://tds.henkel.com/tds5/Studio/ShowPDF/243%20NEW-EN?pid=NCI%207002%20EC&format=MTR&subformat=HYS&language=EN&plant=WERCS, [Accessed 17 September 2021]

[17] Available from: http://www.wzmacniacze-pomiarowe.pl/arch/b1067.pdf, [Accessed 16 September 2021]

[18] Available from: www.ahlborn.com, [Accessed 16 September 2021]

[19] Available from: https://manualsbrain.com/en/manuals/969834/, [Accessed 16 September 2021]

[20] Volf J, Novak V, Ryženko V, Papezova S. Properties of tactile transducer with respect to kind of resistive ink and dimension of electrodes. Measurement: Sensors. 2021;**18**:100311

[21] Force Sensing Resistors Integration Guide and Evaluation Parts Catalog. Camrillo: Interlink Electronics; 2006

[22] Liu H, Meusel P, Hirzinger G. A Tactile Sensing System for the DLR Three Finger Robot Hand. German Aerospace Research Establishment: Institute of Robot and System Dynamics; 1998

Chapter 6

Methodological Aspects of Using Comparators for Metrological Traceability of Instrument Transformers

Valentyn Isaiev and Iurii Anokhin

Abstract

Instrument transformers are widely used in accounting the electricity as well as in protecting the energy generation systems. The accuracy of both voltage and current transformers is a critical parameter in terms of ensuring the reliability of functioning whether high-voltage or low-voltage networks. Two approaches are predominant in characterizing the voltage transformers with high primary rated voltage, these are applying either reference capacitor or reference transformer. Both methods require a device that enables the comparison of either two currents or two voltages. The errors of current transformers are determined by using the means of comparing two secondary currents, one of which is an output of reference transformer and the other is an output of a device under test. The calibration of such comparators may be a very sophisticated procedure. As metrological traceability depends on the measuring instruments and working standards used in calibration, the application of the proposed specific combinations of measuring instruments allows identifying the metrological traceability routes when calibrating the comparators.

Keywords: instrument transformer, current, voltage, calibration, measurement uncertainty, metrological traceability

1. Introduction

The current transformers (CT) are used in both high-voltage and low-voltage networks when the task arises to scale high currents to a value acceptable for measurement by ammeters, electricity meters, and other current-related devices. As for voltage transformers (VT), they are mainly used to scale alternating voltages above 1000 V down to values of 100, 120, 200, 230 V [1]. When it comes to the accuracy of settlement between energy supplier and consumer, the main elements of the measuring circuit are instrument transformers (IT) and meters. This means that the correctness of the accounting of consumed energy resources directly depends on the magnitude of errors of ITs.

The trend of development of modern smart grids and the use of renewable energy sources pushes the technical progress in the field of electrical measurements in the direction of expanding the use of digital low-power current and voltage converters [2–4]. However, today the vast majority of substations in the world continue to operate by using traditional electromagnetic ITs. For this reason, the metrological characterization of ITs does not lose its relevance [5, 6].

Among the varieties of nonconventional approaches to the characterization of ITs, there are those based on the use of a personal computer [7, 8] or with applying the special digitizers [9, 10]. A new approach (by using synchrophasor data) was proposed as an alternative to existing ones to determine the metrological characteristics of transformers [11]. No less important task is the calibration of ITs on-site [12, 13]. However, traditional methods for determining the amplitude and angular errors of ITs are currently used by the vast majority of accredited calibration laboratories because, among the requirements in accreditation, there is the validation of methods and measurement traceability [14]. The conventional methods presented in this chapter are used in the practice of calibration laboratories in the branch of determining the errors of both earthed and unearthed VTs, and CTs.

The reference measuring systems, which may differ both in the reference measuring instrument and in the auxiliary measuring means, are used when calibrating the VT. Consequently, the metrological traceability of the measurement result obtained using the same transformer can reach national standards of different physical quantities. Thus, the measurement method determines the direction (or directions for indirect measurements) of metrological traceability.

A device for comparing (comparator or bridge) the voltage or current of both transformer under test and working standard is almost always used when determining the amplitude component (ratio error) and the angular component (phase displacement) of the error of IT. As for the working standard, the traceability is established quite simply to the primary standard of the National Metrology Institute, but the comparison device is calibrated with an indirect determination of its characteristics. To create a procedure for verifying the correctness of the readout of the comparison device, the metrologist must analyze the sources of uncertainty of such means [15–17]. Several methods for calibrating comparison devices have been developed [18–23], and all they have certain features. This chapter proposes methods for characterizing comparators of almost identical alternating currents or voltages to establish the metrological traceability when using such instruments for the calibration of CTs or VTs.

2. Traditional approaches to characterization of voltage transformers

2.1 The usage of reference voltage transformer

A typical measuring system for calibration of VT using the method of a reference transformer [24] is shown in **Figure 1**.

It could be seen in **Figure 1**, the means of comparing two voltages with a range of values from 20 to 150% (possibly 200%) of the rated secondary voltage of VT with an operating frequency of 50 and/or 60 Hz. Such a comparator can measure the voltage ratio error in the range from 0 to 30%, the voltage phase displacement in the range from 0 to 0.1 rad. Also, the comparator should perform the function of measuring the secondary voltage with a relative uncertainty of $\pm 3\%$, its frequency with an absolute uncertainty of ± 0.05 Hz (below the authors will briefly describe the main differences in the construction of the devices for comparing currents and voltages in the context of the formation of primary measurement information).
A reference high-voltage measure of the voltage transformation ratio is a reference VT with a primary voltage range from 20 to 150% of the rated primary voltage of the VT under test.

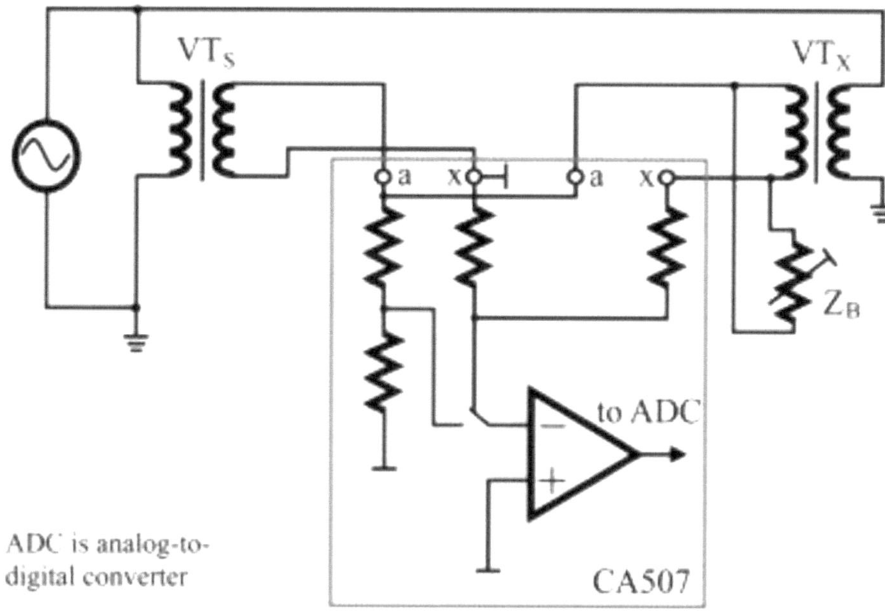

Figure 1.
Measuring system for calibration of VT using the method of a reference transformer.

The VT calibration by the reference transformer is performed with a connection to the comparator using conductors with a resistance of not more than 0.015 Ohm for accuracy class 0.05 and 0.1 or 0.06 Ohm for accuracy class 0.2 and less accurate.

When a message about an incorrect connection appears, the operator needs to change the direction of current flow in the winding of the calibrated VT by swapping the wires at the terminals of the secondary winding. Determination of errors is performed when loading the calibrated VT using the Z_B instrument with a power equal to the rated and quarter of the rated value. Using a high-voltage source, the voltages are set following the requirements of the IEC standard or at the request of the customer. The voltage and frequency can usually be observed on the display of a comparator.

In the case of commensurability of the error of the reference VT and comparator readout, it may make sense to carry out corrective calculations by the formulas:

$$\varepsilon_{VX} = \varepsilon_{VM} + \varepsilon_{VS}; \Delta\varphi_{VX} = \Delta\varphi_{VM} + \Delta\varphi_{VS}. \quad (1)$$

where ε_{VM} is a comparator readout concerning ratio error, $\Delta\varphi_{VM}$ is a comparator readout concerning phase displacement, ε_{VS} is a ratio error of the reference transformer assigned by the calibration certificate, and $\Delta\varphi_{VS}$ is a phase displacement of the reference transformer assigned by the calibration certificate.

For VTs with significant errors, the result read from the display of the comparator can be considered to be the final measurement result.

2.2 The usage of the reference voltage divider

A typical measuring system for VT calibration by the method of reference voltage divider [25] is depicted in **Figure 2**.

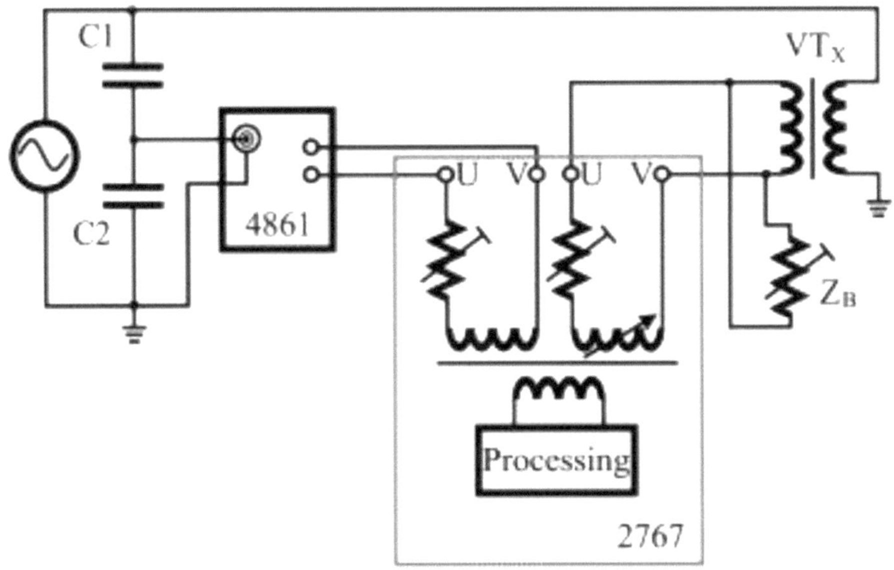

Figure 2.
Measuring system for VT calibration by the method of a reference voltage divider.

The requirements for the reference divider correspond with the requirements for the reference VT specified in the previous section. In this case, the rated scaling factor of the high-voltage standard may not match with the rated scaling factor of the transformer under test.

Figure 2 shows a reference voltage divider consisting of two capacitors C1 and C2. The requirements for conductors should be as described in the previous method. The load device Z_B of the calibrated VT must be set to a value equal to the rated output or a quarter of its value. The high-voltage source should allow setting the voltage required for characterizing the VT.

A comparison device (2767 in **Figure 2** or a specialized volt-phase meter) usually allows controlling the actual voltage and frequency.

In the absence of automatic equilibration as in the design of the 2767 bridge, the value ε_{VM} can be calculated using the formula:

$$\varepsilon_V = 100 \cdot (K_{VX} \cdot U_{X2} – K_{VS} \cdot U_{S2})/K_{VS} \cdot U_{S2}, \quad (2)$$

where U_{X2} and U_{S2} are the secondary voltages measured by the first and the second channels of the volt-phase meter; K_{VX} and K_{VS} are the rated transformation ratios of both the VT under test and the working standard.

2.3 Comparison of design features of comparators of two alternating voltages

In the practice of modern metrological service regarding the calibration of VT, there are two main options for comparing the secondary voltages of the transformer under test and the working standard. The first of them, as noted when considering the method of a reference transformer, requires the use of a standard with an identical transformation ratio. The reference measure should have a large number of primary voltages, which allows providing the full range of primary voltages of the VTs in operation. In this case, it is advisable to use a two-voltage comparator without the possibility of adjusting the output signal of the voltage difference

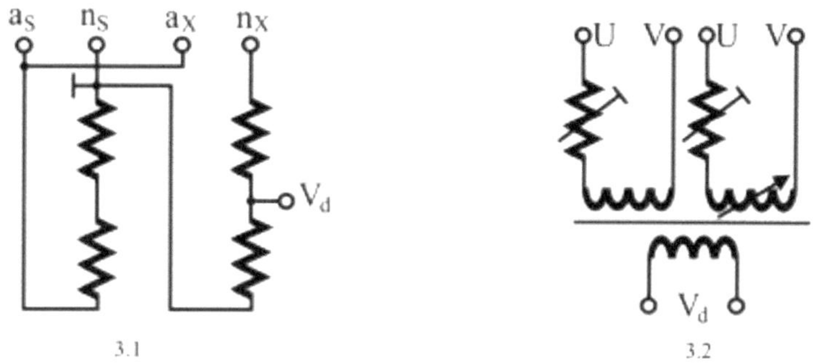

Figure 3.
Input sensor elements of devices for comparing two secondary voltages of VTs.

sensor. **Figure 3** has a position 3.1 with the schematic of the resistive input sensor of the voltage difference of two VTs.

The voltage divider between the terminals a_S and n_S for connecting the secondary winding of the reference VT is required for measuring the actual voltage relative to which the deviation is determined. The voltage divider between the n_S and n_X terminals is used for obtaining information about the potential difference between two sinusoidal signals. Subsequently, both measured signals are transmitted to the phasor measuring analog-to-digital converter (ADC) that allows decomposing the signal into orthogonal components. Further, the processor calculates two values for the ratio error and the phase displacement.

The second option involves the use of a measure with a much smaller number of ratios and the mandatory possibility of balancing (or other methods of comparing significantly different values) of the input voltages in the design of the device for comparing these values. As can be seen from position 3.2 of **Figure 3**, the 2767 measuring bridge has special adjustable active and reactive elements that allow balancing two voltages in a wide range of their ratio (from 0.5 to 10 times). It should be noted that the uncertainty of measurement by this bridge increases as well as the correctness of the readings of such instrument deteriorates with increasing ratio matching factor of voltages of comparable VTs.

It should also be noted that the world market of measuring equipment is saturated with different types of means for comparing two secondary voltages of VTs from different more or less known manufacturers. Some manufacturers use their peculiarities of the input-measuring circuits with or without the possibility of balancing the secondary voltages of the VTs. For example, the AITTS-98 comparator compares two slightly different voltages by using two inductive sensor elements without the ability to adjust the voltage ratio [16].

2.4 The usage of the reference capacitor

The methods for determining the errors of VTs also include the use of two capacitors and a precision transformer bridge built on the principle of current comparison [26]. It is known that the magnetic flux is reduced to the minimum possible value in the transformer core of the current comparator when balancing bridges of similar design [27, 28]. If there is some residual magnetic flux, the proportional signal is generated in the measuring circuit. One of the variants of the bridge measuring scheme with a close inductive coupling is shown in **Figure 4**.

The diagram shows the first stage of determining the errors of the VTs, which determines the basic current ratio of high-voltage and low-voltage capacitors. Currents are generated in two parallel circuits of both reference and unknown

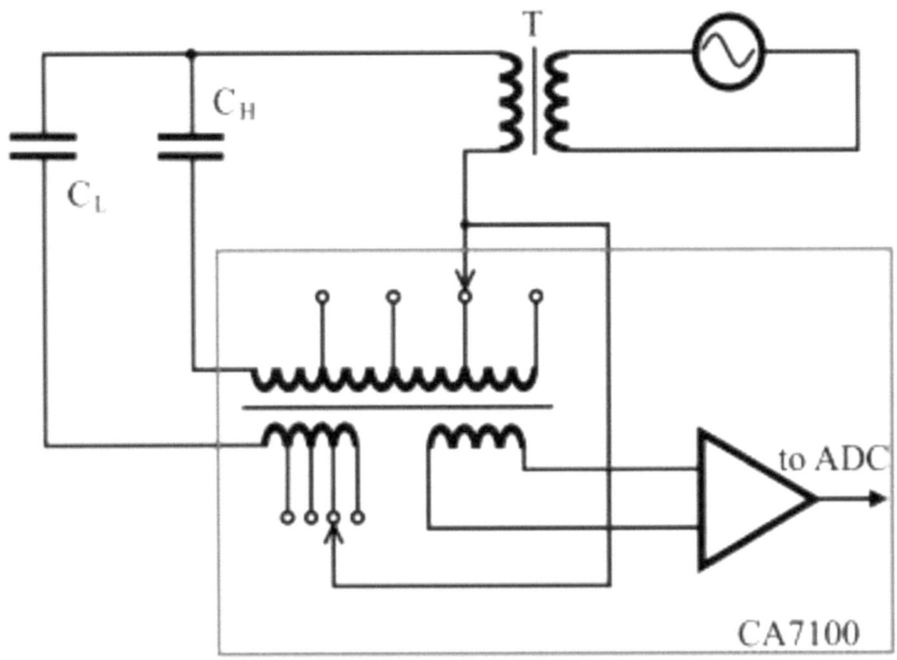

Figure 4.
Measuring diagram of an automatic transformer bridge with two capacitors.

capacitors. The ratio of the turns of the current comparator is automatically adjusted so that the minimum magnetic flux flows through the core. In this case, the ratio of the turns of the current comparator, which minimizes the difference of magnetic fluxes, with some accuracy (which depends on the number of adjustable turns that is a discretization of feasible ratios) will be equal to the ratio of currents flowing in the arms of the bridge with capacitors C_L and C_H. The residual imbalance signal should be amplified, and its additional contribution to the determined current ratio should be calculated by the processor. A typical measuring system for VT calibration by the method of reference capacitor is shown in **Figure 5**.

The principle of balancing the bridge circuit is used as the basis for the measurement by the method of the reference capacitor. Therefore, the most metrologically significant element is a high-voltage transformer bridge, which is a current comparator with a range of compared currents from 0.1 μA to 50 mA. Such a bridge should provide a measurement of both the voltage scaling factor from 0.1 to 10,000 and phase displacement in the range from 0 to 0.1 rad at a frequency of 50 or 60 Hz. The measurement of VT secondary voltage with a relative uncertainty of ±3% and a frequency with an absolute uncertainty of ±0.05 Hz should also be included in the functions of the instrument.

The necessary elements of such a measuring system are two capacitors. Low-voltage electrical capacitor is designed for 10 times voltage of the secondary winding of the VT to be calibrated. C_L is a shielded, highly linear, and highly stable three-electrode electrical capacitor based on film technology and ceramics, with a capacitance in the range from 1000 to 5000 pF depending on the sensitivity of the high-voltage bridge. The tangent of the dielectric loss angle in the operating voltage range should not exceed 10^{-4}.

The high-voltage electrical capacitor must be intended to operate with voltages up to 120% of the VT-rated primary voltage. C_H is a shielded, highly linear, and highly stable three-electrode electric capacitor, made of coaxial electrodes placed in

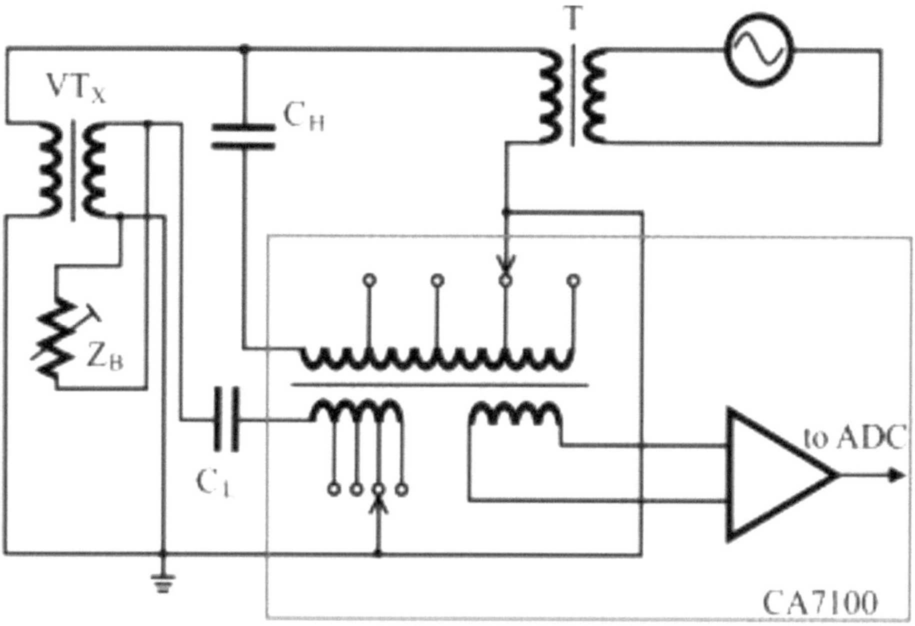

Figure 5.
The second-stage diagram for VT calibration by the method of a reference capacitor.

a case with a gas-insulated dielectric, with a capacitance in the range from 40 to 150 pF depending on the sensitivity of the high-voltage bridge. The tangent of the dielectric loss angle should also not exceed 10^{-4}. When using the method of a reference capacitor, the VT errors are determined from the measurement results obtained during two stages.

At the first stage, it is necessary to determine both the ratio of the currents flowing through two capacitors ($K_{L/H} = I_L/I_H$) and the phase shift angle between these currents ($\varphi_{L/H}$). The voltage across the capacitors should be set at the level of about 1000 V, the bridge should be equilibrated automatically following the command of an operator, and the software must calculate and display certain characteristics.

The measuring circuit at the second stage corresponds to **Figure 5** and differs from the first stage by the presence of a calibrated VT and its load. In this case, the potential input of the low-voltage capacitor is not connected with the similar input of the high-voltage capacitor, but with the secondary winding of the VT under test.

Before turning on the high-voltage source, the burden should be set to the rated or a quarter of the rated value. The high voltage should be set according to the selected calibration points by the reading of the measuring bridge or alternative measuring instrument. At each of the values of the set voltages, the bridge must be balanced as well as both the new ratios ($K_{H/L} = K_{VT} \cdot I_H/I_L$) of the currents flowing through the C_L and C_H capacitors and the new phase shift angles ($\varphi_2 = \varphi_{H/L} + \varphi_{VT}$) must be determined. The ratio error of the VT under test (ε_V) and the phase displacement of voltage ($\Delta\varphi_V$) for each voltage and load set during calibration should be calculated by the following formulas:

$$\varepsilon_V = 100 \cdot \left(1 - K_{L/H} \cdot K_{H/L}/K_{VTr}\right); \Delta\varphi_V = \varphi_{L/H} + \varphi_2, \tag{3}$$

where K_{VTr} is a rated transformation ratio of the VT under test.

3. Metrological characterization of current transformers

3.1 Traditional approaches to characterization of current transformers

Traditionally, the errors of CTs are determined by comparing their secondary current with the secondary current of the reference transformer [29]. A typical measuring system is shown in **Figure 6**.

The current source must allow regulating the current in the primary circuit in the range from 1 to 150% of the rated primary current of the CT under test with a deviation of not more than 10%. The primary current flows through the primary windings of both the working standard and the CT under the test and causes a magnetic flux in the cores of both transformers.

As a result, currents proportional to the primary current, which is the same at each time point, occur in both secondary windings. The burden ZB causes a change in the actual metrological characteristics of the calibrated CT and should allow changing the load in the range from a quarter to the rated power at a power factor of 0.8 with an uncertainty of impedance, which does not exceed $\pm 4\%$. In some cases, it is sensible to use the actual load (or its equivalent) with which the CT operates. The current comparison device must allow the measurements with intrinsic uncertainty in the range from ± 0.03 to $\pm 0.001\%$ when measuring the current ratio error and from $\pm 3.0'$ to $\pm 0.1'$ when measuring the phase displacement depending on the difference between the currents of the working standard and the CT under test.

The amplitude (ε_I) and angular ($\Delta\varphi_I$) errors of the CT should be determined by the differential method following **Figure 6** at the primary current and load by customer order. The connection of devices in the measuring circuit according to **Figure 6** is carried out following the requirements of the operating manual of the current comparison device used. For the CT being tested, the relative current ratio error in percent and the absolute phase displacement in minutes are taken equal to the values at the comparator display.

A variant of implementation of the measurement setup may also be the use of a means for comparing the primary and secondary currents of the CT under test. A feature of such a comparator is the presence of an integrated reference transformer with a variable transformation ratio and high precision. An example of the scheme is shown in **Figure 7**.

The device used for comparing two currents consists of two units (transformer-electronic and electronic-computing) and contains the reference transformer inside. Due to the presence of the integrated standard, such device allows implementing two options for comparison—using the built-in standard or using the external

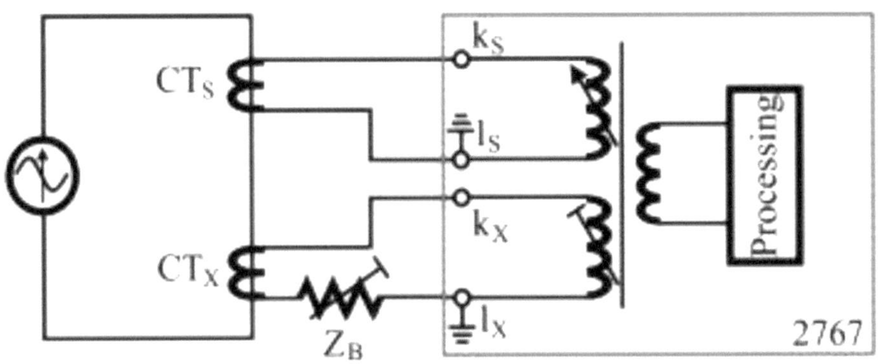

Figure 6.
Measuring system for calibration of CT by means of comparison with a reference transformer.

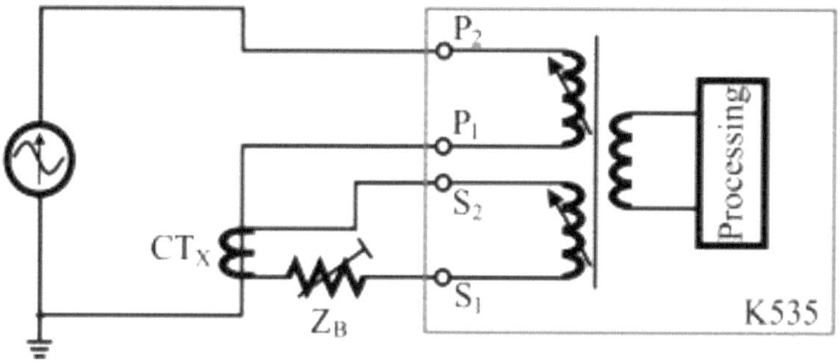

Figure 7.
Measurement setup for calibration of CT with comparing primary and secondary currents.

Figure 8.
Input sensor elements of means for comparing two secondary currents of the CTs.

reference CT. For the transformer under test, this scheme compares the magnetic fluxes generated by both the scaled primary current and the secondary current equaled with the help of reference inductive converters of the transformer-electronic unit. The signal generated by the secondary winding of the magnetic flux comparator is proportional to the difference between the almost equal currents. The second option for using the K535 device requires the presence of two CTs with the same transformation ratios and, therefore, is similar to that described above.

In the general case, the input-measuring circuits of the comparison means create a load effect that can distort the measurement result. Apart from the input inductive current difference sensors, the resistive measuring elements are also used, namely shunts. The resistive and inductive sensors create voltage drops in addition to the load Z_B, and the effect of some comparators on the calibration result was investigated previously [30]. The examples of mentioned sensors are shown in **Figure 8**.

Similar to voltage difference sensors, manufacturers of precision instruments have implemented options of both comparison type and bridge type with the possibility of balancing two currents. Two options presented in **Figures 6** and **7** involve bringing counter-directed magnetic fluxes to almost equal values by selecting the desired number of turns. Item 8.1 of **Figure 8** presents an inductive measuring sensor of the relative difference of currents, which is implemented in the design of the 2767 measuring bridge. By adjusting the number of turns of the input windings, the developer has created the ability to vary the ratio matching factor in the range from 0.5 to 500. The difference flux remaining in the magnetic circuit creates an output information signal proportional to the difference in currents of the working standard and the transformer under test. The comparators CA507 and

HGQA-C are designed for comparing two almost identical currents, which requires the use of a working standard with a transformation ratio equal to the same characteristic of the calibrated CT. Particularly, item 8.2 of **Figure 8** presents a sensor of the relative difference of currents in the design of the CA507 comparator. The left-measuring shunt generates information about the current of the working standard, and the right shunt is intended to measure the absolute phasor difference of two currents. These two measured quantities are decomposed into orthogonal components by using a vector measuring ADC.

3.2 Metrological characterization by measuring current transformer parameters at low voltage

For the needs of operative and mobile determination of CT errors, an instrument for testing the CTs has recently been developed, namely a CT analyzer, which does not require a reference transformer [12, 31, 32]. It is especially convenient when determining the errors of the CTs with big currents located on-site because of the small weight and dimensional indicators. According to the theory of transformer characterization, the equivalent current transformation scheme with some simplifications can be represented as in **Figure 9**.

Each transformer winding has active resistance and reactance, which is shown in **Figure 9** as R_P and X_P for the primary winding and R_S and X_S for the secondary winding. According to the equivalent schematic, the current of the primary circuit is branched and part of it flows through the excitation branch consisting of the active Re and reactive Xe components. Excitation current through this branch is proportional to induced electromotive force E. The CT analyzer must be able to generate both direct and alternating voltages for determining both the active and reactive components of the CT windings and the excitation current [12, 32].

The use of the low-voltage reciprocity principle allows facilitating the measurement procedure in determining the errors when the internal characteristics of the CT are measured. Excitation characteristics, 10% error curve, and composite error are considered in the processing algorithm of the measuring information [12].

Following the user's guide of CT analyzer CT1, the processor of this device calculates some characteristics based on the equations obtained when considering the equivalent circuit. The results are influenced by the electrical resistance of the secondary winding, ambient temperature, etc. The calculated values of the REs are based on an excitation table. The excitation table allows us to find the phase between current and voltage, and the corresponding excitation current due to the given excitation voltage [33].

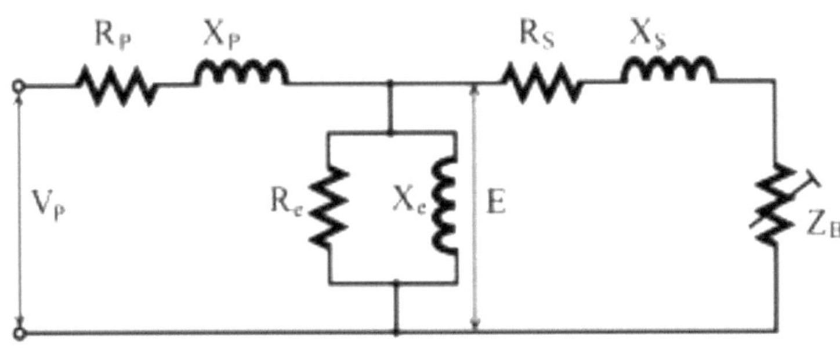

Figure 9.
Equivalent simplified schematic of the CT.

4. Characterization of comparators of two alternating voltages

4.1 Checking the correctness of comparator readings for measuring the ratio error of voltage transformers

The following procedure allows checking the correctness of the readings of the means for comparing two alternating voltages when calibrating the VTs in the range of secondary voltage from 5 to 240 V. When performing the calibration of such comparator, it is necessary to perform the operations of determining the deviation of the amplitude error of the voltage (ε_V) as well as determining the deviation of the phase error of the voltage ($\Delta\varphi_V$).

The measurement setup for calibration of the comparator for determining the ratio errors of VT is shown in **Figure 10**.

The main element of the presented scheme is the P4834 resistance decade box, which allows generating the required voltage ratio error. The P4834 must have terminals for connecting with the intermediate decades, which can be used as a branching point (if this condition is not met, another resistance decade box may be needed). The source of alternating current (AC) generates the required voltage between the first and the last terminals of the P4834. In the measuring scheme, a source of stable-alternating voltage should be used for minimizing the fluctuations of its output voltage to minimize the scattering of the observation results of the two measured signals. The deviation of the voltage amplitude error can be determined by using two precision digital multimeters in the AC voltage measurement mode (voltmeter), such as Agilent 3458A or Fluke 8845A. The right voltmeter is designed to measure the voltage difference between the terminals of the input secondary voltages of both the working standard and the VT under test.

The left voltmeter is required for measuring the voltage at the terminals of the input secondary voltage of the reference VT. The observed voltage difference must be divided by the voltage at the terminals of the input secondary voltage of the reference VT. The resistors R1 and R2 form a divider inside the CA507 comparator, which is designed for measuring the voltage of the working standard. The resistors R3 and R4 also form a divider inside the CA507 comparator, which is designed for measuring the difference between two secondary voltages.

Calibration of the comparator is performed for the input voltage of 50 and 100 V and generated errors are listed in **Table 1**. Since the measurement procedure is the same for each of the specified points, the following description of calibration operations is given for one abstract value of the input voltage and voltage ratio error ε_V.

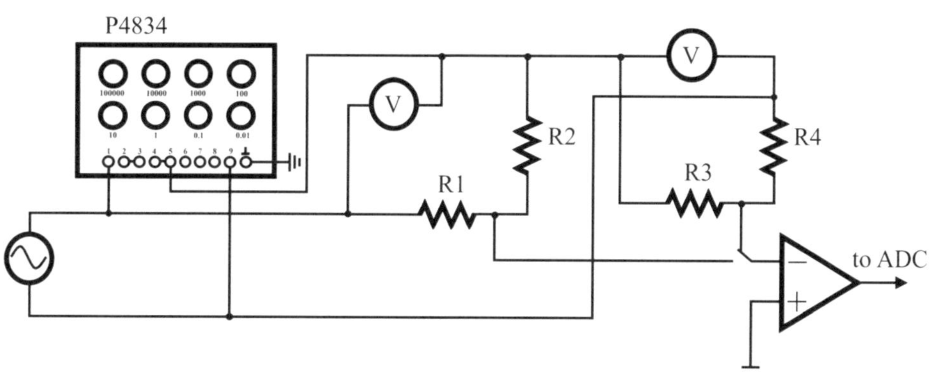

Figure 10.
Measurement setup for calibration of the comparator of two almost identical voltages in measuring the ratio error.

Voltage ratio error (%)	Voltage phase displacement (min)	Resistance of P4830 decade box (Ω)		Capacitance of P5025 decade box (μF)
		For ratio error	For phase displ.	
0.05	—	10,005	—	—
0.1	5	10,010	21,880	100
0.5	15	10,050	7295	100
2.0	50	10,200	21,880	10
5.0	100	10,500	10,939	10

Table 1.
Values of P4830 and P5025 decade boxes for generating the VT errors.

The values measured using voltmeters (the voltage U_S at input terminals for reference VT and voltage U_Δ of a difference between two input voltages) should be committed to memory simultaneously following the command of the operator. The obtained values are stored in the memory of a personal computer and then exported to the electronic protocol. The observations should be repeated the required number of times n obtaining a series of unit values of U_{Si}, $U_{\Delta i}$, and ε_{Vi} ($i = 1, \ldots, n$). The least significant digit Δ_L of the comparator should be considered in evaluating the measurement uncertainty.

The average deviation of the voltage ratio error in units (%) of this metrological characteristic is calculated by the formula:

$$\overline{\Delta_U} = \sum_{i=1}^{n} \left(\varepsilon_{Vi} - 100 \cdot U_{\Delta i}/U_{Si}\right) \Big/ n \tag{4}$$

The combined standard uncertainty for this characteristic can be calculated by the formula:

$$u_\Sigma = \sqrt{\frac{10^4 \cdot \sum_{i=1}^{n}\left(\frac{U_{\Delta i}}{U_{Si}} - \frac{\overline{U_\Delta}}{\overline{U_S}}\right)^2}{n \cdot (n-1)} + \frac{\sum_{i=1}^{n}\left(\varepsilon_{Vi} - \overline{\varepsilon_V}\right)^2}{n \cdot (n-1)} + \frac{\Delta_L^2}{12} + \frac{10^4 \cdot u_\Delta^2}{U_S^2} + \frac{10^4 \cdot u_S^2 \cdot U_\Delta^2}{U_S^4}} \tag{5}$$

where u_Δ and u_S are the standard uncertainties associated with measuring the corresponding voltages U_Δ and U_S.

4.2 Checking the correctness of comparator readings for measuring the phase displacement of voltage transformers

When checking the accuracy of measuring the difference between the phase errors of two VTs, the metrologist should use the scheme shown in **Figure 11**.

The resistance P4834 and capacitance P5025 decade boxes are two important elements that allow generating the required phase shift between two input voltage phasors. The P4834 decade box has a much higher value of electrical resistance than the equivalent resistance created by the P5025 capacitance box. Since the actual phase shift angles, obtained during the calibration of VTs, rarely exceed 100 min, the calculation of reproducible quantity can be carried out by a more convenient formula based on the equality of small angle and its tangent. The alternating voltage source generates the required value at the separated terminals of the P4834 and P5025 decade boxes connected in series. The deviation of the angular voltage error

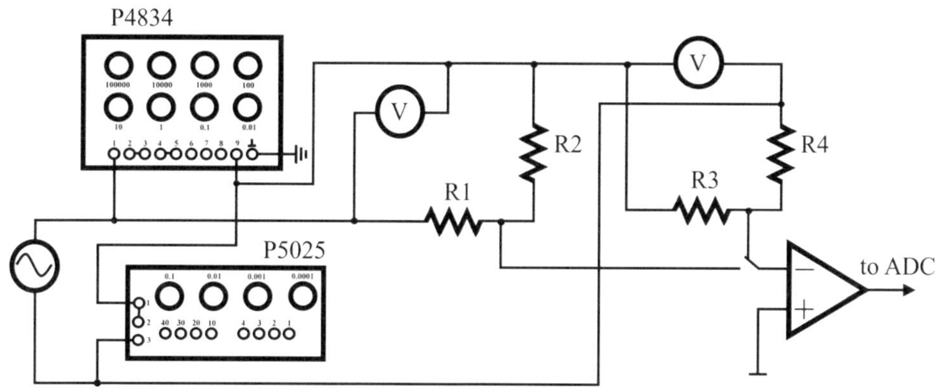

Figure 11.
Measurement setup for calibrating the comparator of two almost identical voltages in measuring the phase displacement.

displayed by the comparator can also be determined by using two precision voltmeters. In **Figure 11**, the right voltmeter is intended for measuring the imaginary component of the voltage present at the comparator terminals for the VT under test. The left voltmeter measures the voltage at the terminals for the reference VT. The obtained value of the imaginary component of the voltage should be divided by the voltage of the reference VT. As for the resistors R1–R4, their function has already been discussed in the previous section.

The calibration of the comparator by phase shift angle should be performed for the values of generated error which are listed in **Table 1**. The following description of calibration operations is given for one abstract value of input voltage and corresponding phase displacement $\Delta\varphi_V$.

The values measured using voltmeters should be committed to memory simultaneously following the command of the operator. The observations should be repeated the required number of times n obtaining a series of unit values of U_{Si}, $U_{\Delta i}$, and $\Delta\varphi_{Vi}$ (i = 1, ..., n). The average deviation of the voltage phase displacement in units (min) of this metrological characteristic is calculated by the formula:

$$\overline{\Delta_{U\varphi}} = \sum\nolimits_{i=1}^{n} \left(\Delta\varphi_{Vi} - 3437.747 \cdot U_{\Delta i}/U_{Si}\right) \Big/ n \tag{6}$$

The combined standard uncertainty for this characteristic can be calculated by the formula:

$$u_\Sigma = \sqrt{3437.747^2 \cdot \left[\frac{\sum_{i=1}^{n}\left(\frac{U_{\Delta i}}{U_{Si}} - \frac{\overline{U_\Delta}}{\overline{U_S}}\right)^2}{n\cdot(n-1)} + \frac{u_\Delta^2}{U_S^2} + \frac{u_S^2 \cdot U_\Delta^2}{U_S^4}\right] + \frac{\sum_{i=1}^{n}\left(\Delta\varphi_{Vi} - \overline{\Delta\varphi_V}\right)^2}{n\cdot(n-1)} + \frac{\Delta_L^2}{12}}. \tag{7}$$

5. Checking the correctness of comparator readings for measuring the errors of current transformers

The process of metrological characterization of the CT can be parted into two stages. The peculiarities of the first stage for determining the small CT errors are outlined in [34].

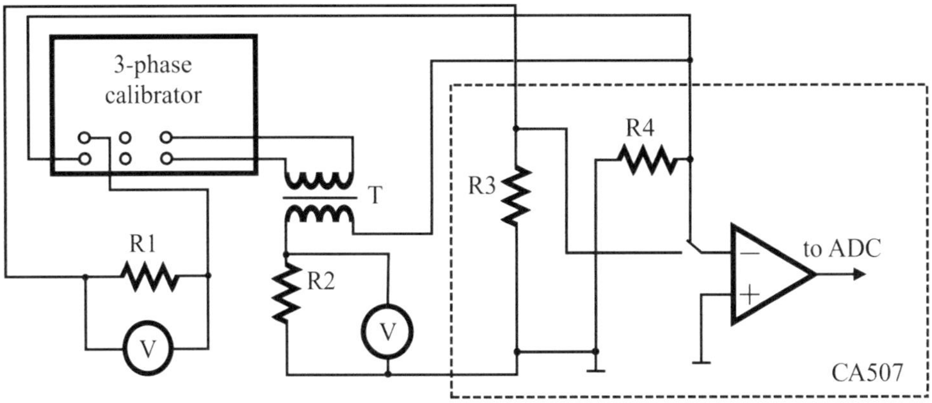

Figure 12.
Measurement setup of comparator calibration when simulating CT errors greater than 0.1%.

Current ratio error (%)	Secondary current I2 of working standard (A)	Current of 3-phase calibrator	
		In phase 1	In phase 3
0.2	1; 5; 6	I2	1.002·I2
0.5		I2	1.005·I2
1.0		I2	1.01·I2
2.0		I2	1.02·I2
5.0		I2	1.05·I2

Table 2.
Currents generated for calibrating comparator when ratio error is larger than 0.1%.

The second stage of metrological characterization of the comparator for larger values of CT errors can be performed by using the scheme of **Figure 12**.

In the second stage, the three-phase calibrator is a source of generated current ratio error (ε_I). The readout deviation of the amplitude error of the current is determined using a voltmeter that measures the voltage U_S at the output terminals of the reference resistance R1, which is in the secondary current circuit of the working standard. The second voltmeter is intended for measuring the voltage U_X at the output terminals of the reference resistance R2 placed in the secondary current circuit of the CT under test. A transformer T with a current ratio of 5A/5A is used for galvanic isolation.

The measuring shunts R3 and R4 in the comparator structure are designed to measure the current of the working standard and the vector difference between the currents of the standard and the transformer under test.

The calibration of the comparator for current ratio error larger than 0.1% is performed for the input currents and generated errors that are specified in **Table 2**. The average deviation of the current ratio error in units (%) of this metrological characteristic is calculated by the formula:

$$\overline{\Delta_I} = \sum_{i=1}^{n} \left[\varepsilon_{Ii} - 100 \cdot \left(K_R \cdot \frac{U_{Xi}}{U_{Si}} - 1 \right) \right] \bigg/ n. \tag{8}$$

where K_R is the relative difference factor, which is determined in characterizing the difference current-to-voltage converter consisting of R1 and R2 when the same current is flowing through both devices; U_{Si} is one observation of left voltmeter readout; U_{Xi} is one observation of right voltmeter readout; ε_{Ii} is one observation of comparator readout.

The combined standard uncertainty for this characteristic can be calculated by the formula:

$$u_{\Sigma} = \sqrt{\frac{10^4 \cdot K_R^2 \cdot \sum_{i=1}^{n}\left(\frac{U_{Xi}}{U_{Si}} - \frac{\overline{U_X}}{U_S}\right)^2}{n \cdot (n-1)} + \frac{\sum_{i=1}^{n}(\varepsilon_{Ii} - \overline{\varepsilon_I})^2}{n \cdot (n-1)} + \frac{\Delta_L^2}{12} + \frac{10^4 \cdot K_R^2 \cdot u_X^2}{U_S^2} + \frac{10^4 \cdot K_R^2 \cdot u_S^2 \cdot U_X^2}{U_S^4}} \quad (9)$$

where u_X and u_S are the standard uncertainties associated with measuring the corresponding voltages U_X and U_S.

6. Investigation of the metrological performance of high-voltage transformer bridge for determining the voltage transformer errors

The method of cyclic permutations is used to conduct experimental studies of the metrological characteristics of the AC transformer bridge [35]. This method allows for determining the deviations of the transformer bridge with high accuracy. Moreover, not the values of capacitance but their ratios are reproduced, and for dielectric loss tangents, their difference is reproduced. To apply this method, a group of capacitors in the amount of $(m + n)$ is used, which is shown in **Figure 13**.

The presented schematic has some simplifications and does not contain all the elements of a special impedance box for characterizing the AC bridge. With the help of a group of switches, which are not depicted, the necessary ratios between the total capacitances of n-Section and m-Section are formed. To determine the deviation of the ratio of currents m/n, the metrologist has to perform $(m + n)$ sequential measurements using $(m + n)$ capacitors. In each measurement, n capacitors are connected in parallel to the C_0 input of the AC bridge, and m capacitances are connected to the C_X input. In each subsequent measurement, one of the capacitors from the C_0 input is switched to the C_X input, and one of the capacitors from the C_X input is switched to the C_0 input.

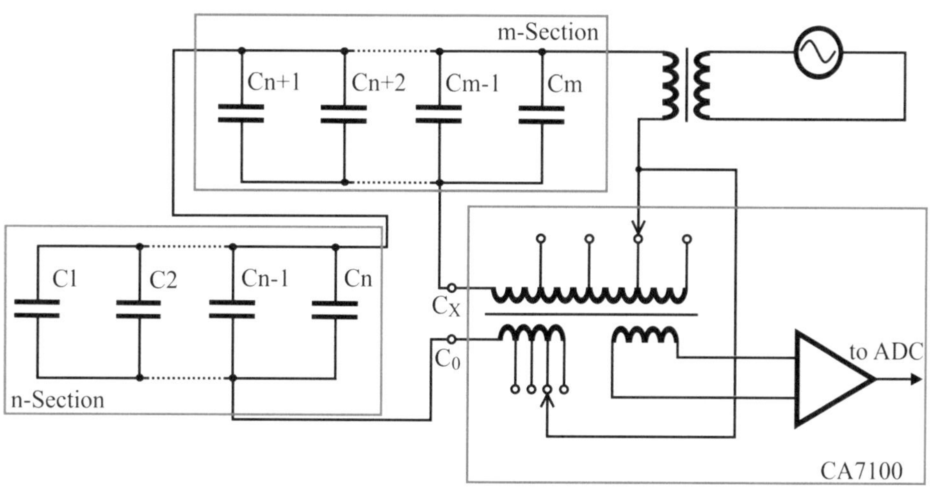

Figure 13.
Measurement setup for characterization of transformer bridge.

During $(m + n)$ measurements, each of the capacitors is connected m times to the input C_X and n times to the input C_0. Processing the measurement results, the metrologist has to determine the relative deviation in measuring the capacitance by the formula

$$\delta_{Cm/n} = \sum_{i=1}^{m+n} \frac{C_{Xi}/C_0 - m/n}{m/n} \Bigg/ (m+n) \tag{10}$$

and the absolute deviation in measuring the dielectric loss tangent by the formula

$$\Delta tg\delta_{m/n} = \sum_{i=1}^{m+n} \Delta tg\delta_i \Big/ (m+n) \tag{11}$$

where $\Delta tg\delta_i$ is the absolute deviation in measuring the dielectric loss tangent at ith stage of the measurement cycle.

7. Routes of metrological traceability in instrument transformer calibration

Figure 14 shows a diagram of the route of traceability of the measurement result obtained by VT calibrated by the method of reference transformer. As for the reference VT (RVT), it should be calibrated most often using the national standard (NSVSC) of the high-alternating voltage scaling coefficient [36]. This traceability route connects the VT under test with mentioned national standard through the chain of calibrations when both the voltage ratio error ε_V and phase displacement $\Delta\varphi_V$ are determined (boundary branches for the ratio error and phase displacement in **Figure 14**).

The comparator (VTC) of two almost identical voltages, calibrated by the procedure described in this chapter, provides traceability to the national standard of the electric voltage. This becomes clear when considering **Figures 10** and **11**, where two voltmeters are the working standards (two middle branches in **Figure 14**).

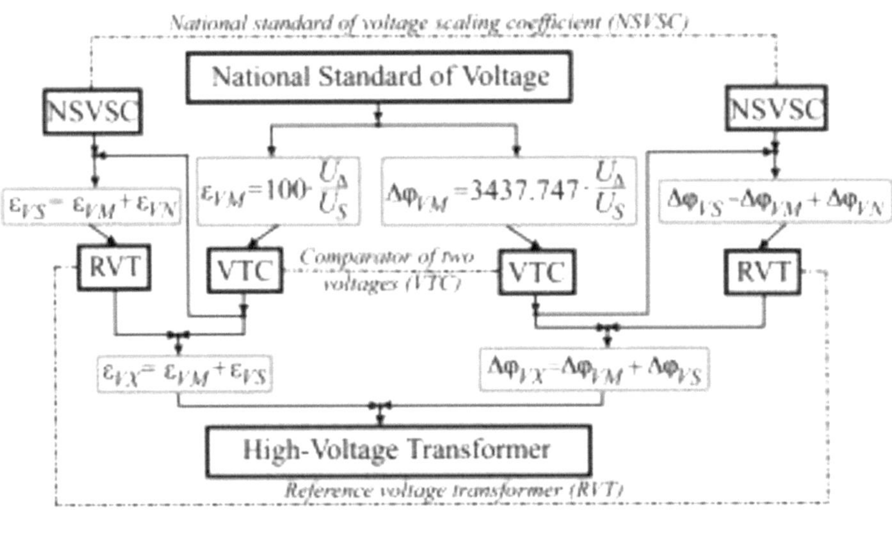

Figure 14.
Traceability routes for the measurement result with VT tested using the reference transformer.

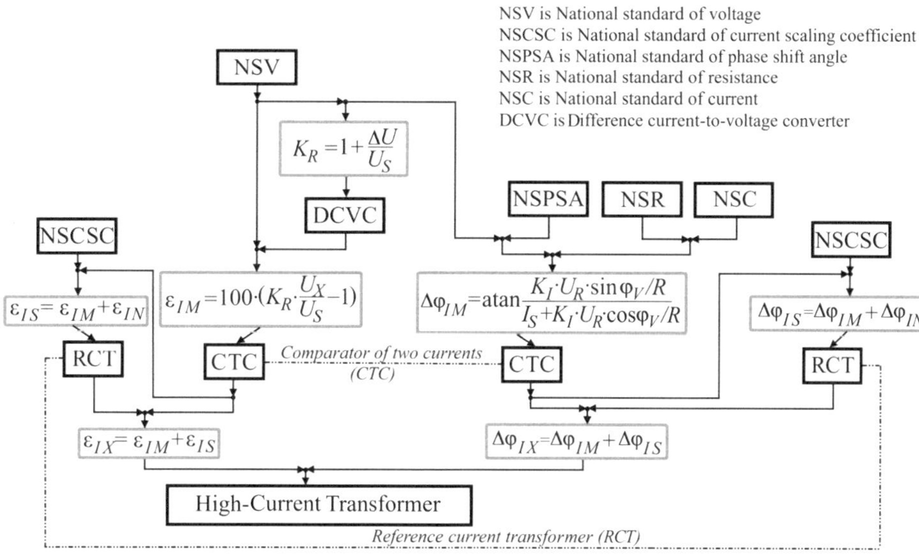

Figure 15.
Traceability routes for the measurement result with CT calibrated by the conventional method.

Regarding CTs, this chapter describes the method used by all National Metrology Institutes in the world when calibrating CTs with primary current up to 10 kA. As in the method of reference VT, a working standard and a comparator are used (see **Figure 6**).

Figure 15 shows a diagram of the traceability route of the measurement result obtained by using the calibrated CT. As for the reference CT (RCT), it should be calibrated most often using the national standard (NSCSC) of the scaling coefficient of high-alternating current [36]. This traceability route connects the CT under test with the last-mentioned national standard through the chain of calibrations when the current ratio error (ε_I) and phase displacement ($\Delta\varphi_I$) are determined (boundary branches for the ratio error and phase displacement in **Figure 15**).

The comparator (CTC) of two almost identical currents, calibrated by the procedure described in this chapter, provides traceability to several national standards. Considering measurement setup for determining the low CT errors [34], it is seen that the phase meter measures the phase shift angle φ_V between two voltages that is traceable to the corresponding national standard (NSPSA) of the phase shift angle. The readout U_R of the voltmeter should be compared with the value of the national standard (NSV) of voltage as well as the reference resistance R with the value of the national standard (NSR) of resistance, and the readout I_S of the ammeter should be compared with the value of the national standard (NSC) of electric current. The current branching factor K_I depends only on the ratio between voltmeter input impedance and reference resistance, which should be characterized at power frequency [37].

When considering **Figure 12**, one can see two voltmeters and a differential current-to-voltage converter (DCVC) consisting of the resistors R1 and R2. An electric current of the same magnitude flows in the circuit of both R1 and R2 when calibrating the last instrument, causing the next formation of the expression for determining the relative resistance difference $\Delta R/R = K_R - 1 = \Delta U/U_S$. Thus, the task of calibrating such an instrument is reduced to determine the relative voltage difference that occurs at the output terminals of the reference resistances.

Given the aforesaid, it is obvious that the metrological traceability reaches the national standard NSV when applying the comparator by the approach described.

8. Conclusion

Often high-voltage electrical networks provide the transfer of energy resources outside the state and the accounting chain must contain instrument transformers that are trusted. The materials presented in this chapter help to ensure the necessary trust through the application of the procedures with clearly defined routes of metrological traceability.

The method for calibrating the voltage transformer using the reference transformer gives traceability up to the national standard of the high-alternating voltage scaling coefficient. The application of comparator, characterized by the proposed method, sets a connection with the national standard of the electric voltage for both the voltage ratio error and the phase displacement.

The method for calibrating the current transformer using a reference transformer gives traceability to the national standard of the alternating current scaling coefficient. The application of comparator, characterized by the proposed method, sets connection with the national standard of the electric voltage for the current ratio error. When measuring the small phase displacement precisely, the traceability is provided up to several national standards, namely, the standard of the phase shift angle, the standard of electric voltage, the standard of electric resistance, and the standard of electric current.

Author details

Valentyn Isaiev and Iurii Anokhin*
State Enterprise "Ukrmetrteststandard", Kyiv, Ukraine

*Address all correspondence to: y.l.anokhin@gmail.com

References

[1] IEC 61869-3:2011. Instrument Transformers—Part 3: Additional Requirements for Inductive Voltage Transformers. Geneva: IEC; 2011. p. 60

[2] Schmid J, Kunde K. Application of non conventional voltage and currents sensors in high voltage transmission and distribution systems. In: Proceedings of the IEEE International Conference on Smart Measurements of Future Grids (SMFG); 14-16 November 2011; Bologna. New York: IEEE; 2012. pp. 64–68

[3] ENG61 FutureGrid. Final Publishable JRP Report. Non-Conventional Voltage and Current Sensors for Future Power Grids. European Union: EURAMET; 2017. p. 34

[4] Crotti G, Gallo D, Giordano D, Landi C, Luiso M, Cherbaucich C, et al. Low cost measurement equipment for the accurate calibration of voltage and current transducers. In: Proceedings of the IEEE International Instrumentation and Measurement Technology Conference (I2MTC); 12-15 May 2014; Montevideo. New York: IEEE; 2014. pp. 202–206

[5] Alanazi SS, Alayli AR, Alrumie RA, Alrobaish AM, Ayhan B, Çayci H. Establishment of an instrument transformer calibration system at SASO NMCC. In: Proceedings of the Conference on Precision Electromagnetic Measurements (CPEM 2018); 8-13 July 2018; Paris. New York: IEEE; 2018. pp. 1–2

[6] Slomovitz D, Santos A, Sandler R, Barreto G, Aristoy G, Casais J, et al. SIM comparison of ac current ratio using instrument current transformers. In: Proceedings of the Conference on Precision Electromagnetic Measurements (CPEM); 24-28 August 2020; Denver. New York: IEEE; 2020. pp. 1–2

[7] Jagadeesh Kumar V, Sunil C, Sankaran P. PC-based method for the measurement of instrument transformer errors. In: Proceedings of the 11th IMEKO TC-4 Symposium. Trends in Electrical Measurement and Instrumentation; 13-14 September 2001; Lisbon. Budapest: BME MIT; 2001. pp. 74–77

[8] Mohan NM, George B, Kumar VJ. Virtual instrument for testing of current and voltage transformers. In: Proceedings of the IEEE Instrumentation and Measurement Technology Conference Proceedings; 24-27 April 2006; Sorento. New York: IEEE; 2007. pp. 1163–1166

[9] van den Brom HE, Rietveld G, So E. Sampling current ratio measurement system for calibration of current transducers up to 10 kA with 5 ppm uncertainty. IEEE Transactions on Instrumentation and Measurement. 2015;**64**(6):1685–1691. DOI: 10.1109/CPEM.2014.6898232

[10] George N, Vooka P, Sai Kartheek Bandi A, Gopalakrishna S. A novel dual slope conversion technique for measurement of ratio and phase errors of current transformer using comparison method of testing. Measurement: Journal of the International Measurement Confederation. 2021;**179**:109458. DOI: 10.1016/j.measurement.2021.109458

[11] Hinge TP, Dambhare SS. A novel approach for calibration of instrument transformers using synchrophasors. In: Proceedings of the National Power Systems Conference (NPSC); 19-21 December 2016; Bhubaneswar. New York: IEEE; 2017. pp. 1–5

[12] Yang T, Zhang G, Hu X. System design of current transformer accuracy tester based on ARM. In: Proceedings of the IEEE 8th Conference on Industrial

Electronics and Applications (ICIEA); 19-21 June 2013; Melbourne. New York: IEEE; 2013. pp. 634–639

[13] Mohns E, Latzel HG, Wittig C. Power comparator based on-site calibration of isolating current transformers. MAPAN. 2009;**24**:67–72. DOI: 10.1007/s12647-009-0008-8

[14] ISO/IEC 17025:2017. General Requirements for the Competence of Testing and Calibration Laboratories. Geneva: ISO/IEC; 2017. p. 32

[15] Muscas C, Peretto L, Sasdelli R, Briani A. An automatic test equipment for the calibration of voltage transducers. IEEE Transactions on Instrumentation and Measurement. 2001;**50**(6):1638–1643. DOI: 10.1109/19.982959

[16] Iwanusiw OW. Microprocessor-based automatic instrument transformer comparator. IEEE Transactions on Instrumentation and Measurement. 1983;**32**(1):165–169. DOI: 10.1109/TIM.1983.4315033

[17] Betts PJ, Baghurst AH, Hansom DS. Self-calibratable voltage transformer testing set. IEEE Transactions on Instrumentation and Measurement. 1999;**48**(5):906–908. DOI: 10.1109/19.799645

[18] Lee YS, Jung JK, Kim KT. Fabrication of a current transformer test set calibrator and its applications. Measurement Science and Technology. 2015;**26**:085015. DOI: 10.1088/0957-0233/26/8/085015

[19] Çayci H. A complex current ratio device for the calibration of current transformer test sets. Metrology and Measurement Systems. 2011;**18**(1): 159–164. DOI: 10.2478/v10178-011-0015-2

[20] Isaiev V. Method of reference values defining for calibration of two alternating currents comparator with using oscilloscope. In: Proceedings of the IV International Scientific and Practical Conference: Methodology of Modern Research; 31 March 2018; Dubai. Warsaw: RS Global; 2018. pp. 42–49

[21] Çayci H. Application of digital sampling method for voltage transformer test set calibrations. In: Proceedings of the 17th IMEKO TC-4 International Symposium, 3rd Symposium IMEKO TC-19 and 15th IWADC Workshop Instrumentation for the ICT Era; 8-10 September 2010; Kosice. Budapest: IMEKO; 2010. pp. 1–5

[22] Jung JK, Faisal A, Lee YS, Kim K. Calibration of voltage transformer test set by employing voltage divider. In: Proceedings of the 29th Conference on Precision Electromagnetic Measurements (CPEM 2014); 24-29 August 2014; Rio de Janeiro. New York: IEEE; 2014. pp. 382–383

[23] Isaiev V, Velychko O. Precise low-cost method for checking accuracy of current transformers calibration unit. In: Proceedings of the 24th IMEKO TC-4 International Symposium; 14-16 September; Palermo. Budapest: IMEKO; 2020. pp. 406–410

[24] Wang X, Mohns E, Jang CY, Raether P, Yang H. Bilateral comparison of voltage transformer measuring systems of NCHVM and PTB. In: Proceedings of the 23th IMEKO TC-4 International Symposium; 17-20 September 2019; Xi'an. Budapest: IMEKO; 2019. pp. 57–62

[25] Styblikova R, Draxler K, Kikalo V, Kopshyn V. Interlaboratory comparison of ac voltage ratio standards in the range up to 35 kV. In: Proceedings of the IEEE International Instrumentation and Measurement Technology Conference; 10-11 May 2011; Hangzhou. New York: IEEE; 2011. pp. 1–6

[26] Jung JK, So E, Park YT, Kim M. KRISS-NRC intercomparisons of calibration systems for instrument transformers with many different ratios at power frequency. IEEE Transactions on Instrumentation and Measurement. 2009;**58**(4):1023–1028. DOI: 10.1109/IMTC.2011.5944292

[27] Kusters NL, Petersons O. A transformer-ratio-arm bridge for high-voltage capacitance measurements. IEEE Transactions on Communication and Electronics. 1963;**82**(5):606–611. DOI: 10.1109/TCE.1963.6373272

[28] Petersons O, Anderson WE. A wide-range high-voltage capacitance bridge with one ppm accuracy. IEEE Transactions on Instrumentation and Measurement. 1975;**24**(4):336–344. DOI: 10.1109/TIM.1975.4314455

[29] Dimitrov E, Kumanova G, Stybliková R, Draxler K, Dierikx E. Final report EURAMET.EM-S30 on EURAMET project 1081: Supplementary comparison of measurements of current transformers. Metrologia. 2010;**47**: 01001. DOI: 10.1088/0026-1394/47/1a/01001

[30] Isaiev V, Velychko O, Anokhin Y. Comparator effect on equivalence of results of calibrating current transformers. Eastern-European Journal of Enterprise Technologies. 2019;**5**(101): 6–15. DOI: 10.15587/1729-4061.2019.177415

[31] Wang Y, Liu T, Hu X. Study of current transformer calibrating system based on equivalent model. In: Proceedings of the IEEE 10th International Conference on Industrial Informatics; 25-27 July 2012; Beijing. New York: IEEE; 2012. pp. 886–890

[32] Yanping Z, Wei G, Jianwei W, Xinghui D. Design and implementation of the CT analyzer on the basis of the low pressure test principle. Procedia Engineering. 2017;**174**:1387–1392. DOI: 10.1016/j.proeng.2017.01.295

[33] Omicron. CT analyzer user manual [Internet]. 2008. Available from: http://userequip.com/files/specs/6031/CT-Analyzer_user%20manual.pdf [Accessed: 06 September 2021]

[34] Isaiev V, Velychko O. Metrological characterisation of current transformers calibration unit for accurate measurement. ACTA IMEKO. 2021; **10**(2):6–13. DOI: 10.21014/acta_imeko.v10i2.918

[35] Cutkosky RD, Shields JQ. The precision measurement of transformer ratios. IRE Transactions on Instrumentation. 1960;**I-9**(2):243–250. DOI: 10.1109/IRE-I.1960.5006925

[36] KCDB. Calibration and measurement capabilities—CMCs [Internet]. 2021. Available from: https://www.bipm.org/kcdb/ [Accessed: 06 September 2021]

[37] Isaiev V, Nosko S. Power frequency characterization of resistance decade box for calibrating ac comparator. International Journal of Scientific and Engineering Research. 2019;**10**(10): 1450–1456